The Joys of Beekeeping

The Joys of Beekeeping

© Richard Taylor

ISBN 978-1-908904-08-9

Published by Northern Bee Books, 2012
Scout Bottom Farm
Mytholmroyd
Hebden Bridge HX7 5JS (UK)

Printed by Lightning Source, UK

The Joys of Beekeeping

Richard Taylor, 1919-2003, really had two parts to his interesting life. He received his PhD in Philosophy and became a professor at three prestigious universities in the United States and a visiting lecturer at others. In that field he was a noted author of several books and numerous papers. The other part of his life was devoted to beekeeping. He was well known for producing prize-winning comb honey and was much in demand to give presentations on both beekeeping in general and his method of producing comb honey. At beekeeper meetings he was always glad to share his knowledge of honey bees and also how he produced his famous comb honey. Richard began writing articles for *Gleanings in Bee Culture*, today known as *Bee Culture*, in 1966. His regular monthly articles began in 1970 and continued until 2000. Two of his beekeeping books, *The How-to-Do-It Book of Beekeeping* and *The New Comb Honey Book*, are out-of-print, but can be found in the used book market. However, his other two books, *Joys of Beekeeping* and *Beeswax Molding and Candle Making*, are available. Over the years some people became beekeepers after reading Joys; all who read this book find a new enjoyment in keeping bees.

Anne Harman

Contents

Preface

Books on apiculture describe how to produce honey but they neglect to note how beautiful it is. They explain how to tend bees but they do not say how joyous this can be. They describe swarms but say nothing of how inspiring it is to behold one. It is omissions like these that I have tried to rectify. I have tried to give an intimation of the sources of happiness that beekeepers have discovered and to convey my own philosophy that is so closely interwoven with my beekeeping. I have also described how. I do certain things connected with bees.

In doing so, I have had in mind readers who have never seen a beehive and who have no intention of ever owning one but who might want to know how things are done by devotées of this strange craft. I have also had in mind my fellow victims of this obsession, hoping that I have expressed something of their own joys and supplied them with a new idea or two.

Some of the things I have written have appeared in the pages of *Gleanings in Bee Culture* as part of my column "Bee Talk."

I am grateful to the editors for letting me use the material again here.

PREFACE TO THIS EDITION

Soon after my book appeared ten years ago the price of honey, together with everything else, rose rapidly. Within about a year it had doubled, infusing new life and excitement into beekeeping. But this dramatic change rendered obsolete and misleading the estimates and figures I had given in my last chapter, which deals with keeping bees for a living. I have accordingly revised the crucial parts of that chapter.

Otherwise the book, like most things pertaining to apiculture, remains the same.

During those ten years I received, from time to time, reports of people who, seeking some kind of new life for themselves, thought they might have discovered it from reading my book. Some even abandoned what they had been doing until then, to give themselves wholeheartedly to the art of beekeeping. I had not anticipated this, and it has sometimes made me feel uncomfortably responsible. Had I perhaps romanticized the beekeeper's life? Had I led people into what might turn out to be disappointing, perhaps fatefully so? Thinking about this I have, however, concluded that my book expresses, honestly and without embellishment, the joys that I have found tending my bees. If others find the same rewards then I rejoice. And if anyone, seeking those joys, finds less than he had anticipated, then I can only plead that no more was promised. I think beekeeping is not for everyone. Indeed, its fulfillments are reserved for a relative few. It would be very sad if any of these failed to discover them. So my book is a call to these.

20 April 1984

ONE

The Taste of Joy

My book is a rejoicing, and I have no other objective in writing it. It is not primarily meant to be instructive, except indirectly with respect to that most elusive of goals, the achievement of happiness. I can write of this because I have tasted it or indeed, because I have been expressing it in countless ways most of my life. This is not a boast but a thanksgiving, for there is nothing here for which I can credit myself. I love nature, and honeybees are among the most exquisite and inspiring expressions of nature. The life that the husbandry of them has styled for me is the source of my rejoicing.

My bees have not made me rich. Had it been otherwise, had this way of life been a source of wealth, as business and speculation sometimes are, then I would have had my reward, but that reward would not have been happiness. Yet the knowledge that I could have depended upon my bees for at least a meager livelihood has given me a sense of independence, which is itself a joy.

Such joy is no passive delight. Keeping bees in a serious way, with concentration and art, is toilsome, sometimes exhausting, frustrating and discouraging. But it is also made the more joyous for the overcoming of all this. The image of happiness is not that of the patron of an amusement park, nor is it that of someone burdened with galling work, nor that of greed measuring up its gold. Part of the image, at least, is that of a challenge met, a purpose achieved—and achieved with great effort.

The basic reason for any pursuit is to find happiness. Many persons seek it through wealth, power or prestige, and while some of them do find these things, it is doubtful whether they ever find more than the most specious happiness. The ancients, who thought more deeply about happiness than we do, were unanimous in rejecting these goals as sources of it. They thought that happiness consisted of having a good demon, so they called it eudaemonia. There is no doubt that the honeybee has been my demon, and an immeasurably good one as long as I can remember. Happiness cannot depend upon the gifts of our fellows, nor upon their approval, for what they bestow today they can as easily withdraw tomorrow.

We are all the creation of the same nature, or as some perceive it, of God, and we share this lovely earth with a multitude of things great and small. We were not given the world to dominate it, to subdue it or exploit it as though it were a kind of warehouse placed at our disposal and for our exclusive benefit. Rather, we were given it to make our home in it, to share it, to glorify it and to glory in it.

One's happiness is, of course, something personal, something more his own than any possession. We do not all find it in the same way. Some never find it at all. Possibly most never do, even when the means are at hand. But I have found my bees and all the countless things I associate with them a constant and unfailing source of it. I know that not all persons are of this temperament. Some look upon this obsession of mine with incomprehension, some with amusement or curiosity, while I in turn pity them, for they lack the capacity for these particular joys so fulfilling to me.

I can only thank God that, in creating this bounteous earth, He included these tiny creatures whose organization still defies the sublimest intellects and whose labors yield the golden honey that has been prized by human beings since before recorded history. Without bees my own existence would be a shadowy thing, like a world without flowers or without stars or without the songs of birds. The world of men is always uncertain, seldom inspiring, often a source of discouragement and dismay. But the keeper of bees, like anyone who has welded his life to the cycles and patterns of nature, can always turn to his tiny creatures and his craft. The bees have a perpetual store of surprises in readiness for their owner, but they also have a constancy beneath it all. No doubt, there are many fountainheads of happiness, but none equals those nature has provided from the beginning of time. Certainly human artifices, even those we have been taught to consider great, are nothing in comparison. They are not needed, in fact, are dismissed as encumbrances, by those who have tasted the changeless serenity that nature so freely offers.

Even as a boy, driven by the passions and impulses that make youth so tumultuous and blind and filled with folly, I noted this serenity in beekeepers. From time to time I would see one sitting out in his battered chair, basking in the peace and sweetness of the setting, surrounded by his hives of bees, his orchard, his small parcel of land, his quiet and ordered world. Overhead soared the purple martins belonging to the large apartment birdhouse he had erected for them on his barn roof, while around him, in spring, the cherries bloomed and the bees hummed, just as they had for a million springs gone by, and as they will for more millions to come.

There was another beekeeper who was old and had lost some of his strength. I occasionally helped him in his bee yards. One of these yards was set in a beautiful grove of locust trees on the side of a hill, outside of town, and we sometimes went there together in his ancient car, an open Model T Ford, which in those days could be purchased in running condition for ten dollars. I was surprised then how happy I was, riding along in the open air with this old man, an object of curiosity to my friends, giving him a hand, and in turn learning about bees. I had not expected this reward of inner happiness, but having found it I never forgot it.

In the meantime, I had my own hive of bees, purchased for four and a half dollars. Soon I had two. The second was the crude product of one of my early attempts at carpentry. I sold the

comb honey to my neighbors for fifteen cents each, a good price in those days, and I have never forgotten the beauty and goodness of that honey made from the sweet clover that then bloomed everywhere. When my studies, followed by military service, forced me to abandon for a time the small beginning I had made in apiculture, I nevertheless preserved my memories of it. In whatever part of the world I found myself, my thoughts went back to my bees, to the fruit trees in full bloom and the beehives beneath them, to the old man who taught me and to his apiary set out in the grove of black locusts, to the soft maple in the spring ringing to the hum of the bees, and all the numberless associations of those golden hours.

I have often thought since then that anyone who does not taste happiness in his youth is in danger of never finding it. Those who do find it, on the other hand, are likely to return to those sources of it, sooner or later, as I certainly did. How blessed I am to have grown up in a village where bees were kept, to have known beekeepers who taught me about them and to have had a loving family to encourage my interest in them. I believe this has been the constant source of the most lasting joys of my lifetime, and my existence would have been indescribably poorer without it.

TWO

The Language of Beekeeping

My book is written for all who love nature—for those who have tended bees most of their lives as well as those who have never seen a beehive and who know nothing of beekeeping except what they have gathered from enjoying honey occasionally, sometimes seeing a bee on a flower and perhaps at some time or other receiving a sting. For these I am going to briefly explain the words we beekeepers use, thereby saving the need to explain them as they come up. This language is not esoteric. It only seems so to those whose ears are unfamiliar with it.

To begin with then, a *hive* is a man-made boxlike container in which bees live and establish their colony, sometimes inappropriately called their "nest." The *colony* is a social unit dwelling within the hive (or, in nature, within a hollow tree) and consisting normally of one *queen*, or female, several hundred *drones*, or males, and thousands upon thousands of *workers*, appropriately so-called, who perform every task of the colony except reproduction.

My *apiaries* are groups of beehives, sometimes twenty or more, at a given place. I also call them my *bee yards*, or simply

9

yards. I do not, with the exception of my home yard, actually own these places, although of course I own the hives that are kept there. Instead, I have persuaded farmers to let me use out-of-the-way nooks and corners of their land, which they are happy to do in return for pollination and some free honey from time to time.

Now for the hive itself. A *frame* is a wooden rectangular device within which the bees are induced to build *comb*. Normally a single-story hive contains ten such frames arranged side by side, and hence ten combs. The more typical two-story hive contains twice as many frames. The comb is a complex waxen structure, built by the worker bees and composed of thousands of hexagonal cells on both sides. Since a frame is removable, I can at will remove combs, shift them from hive to hive, spin the honey from them, do whatever I wish with them. This fact constitutes the basis of modern apiculture.

A *brood frame* is one containing a comb that is used by the bees for raising their young. The queen lays eggs therein, one egg to each cell of the comb, and these hatch into tiny grubs that soon metamorphose into adult bees. An *extracting frame* is a frame containing a comb used by the bees for honey storage, and hence one from which I spin (or "extract") delectable honey that then appears on my customers' tables. Brood frames and extracting frames are identical, except that the latter are often smaller. By means of an unbelievably clever ruse, I can induce the bees to build their combs in the frames where I want them built, rather than where nature would guide them: I insert in each frame a sheet of pure beeswax called *foundation*, which has the pattern of the honeycomb embossed on each side. The bees, coming upon this sheet of foundation, take it to be the beginning of a honeycomb. They simply complete it, or as beekeepers say, "draw it out." I do not actually make this foundation, and hence can claim no credit for its ingenuity. Instead, I deliver beeswax to the bee supply factory and for a small "work up" fee, they convert it to foundation for me.

A hive consists of several stories, normally two at the beginning of the season, then more as greater quantities of honey come to be stored there. The bottom story is called the *brood chamber*, for here is where the queen normally lays thousands upon thousands of eggs, in the brood frames. The second story is called the *food chamber*. It is identical to the brood chamber, and is used for much the same purpose, except that in the fall it is mostly filled with honey upon which the bees subsist through the winter; hence its name. These two stories make up the beehive proper.

All stories above these two are called *supers*, since they are the stories superimposed upon the hive. These contain only extracting frames. Here is where the bees store all the *surplus* honey, which, at the end of the season, I harvest. In a normal year I take about one hundred pounds of honey from each hive, sometimes less, but sometimes much more.

Beeswax is exuded by the workers from special glands as tiny white flakes and then sculpted and molded by them into the exquisite and complex geometric pattern of the honeycomb. It is, like honey, ultimately derived from the nectar of flowers, although it bears no other resemblance to honey. It is unique among waxes for its hardness and beauty and is an extremely valuable byproduct of beekeeping.

Propolis is a fragrant, gumlike substance gathered by the bees from various trees and plants and used by them to caulk holes and cracks. The word is Greek and means literally "before the city." The ancients apparently believed that this substance is used by the bees as the foundation of their city, or *polis*.

A *queen* is a female honeybee and the mother of the entire colony. For most of recorded history she was called the king bee, after human analogy, until it was finally noticed that she laid eggs. Normally there is only one queen in a hive. She is created by the workers from the tiny grub that would normally become only another worker bee. Nothing is understood of the organic chemistry of this remarkable transformation, except that it is achieved by a hormonal secretion from the pharyngeal glands of the workers, called "royal jelly," with which the developing larva is lavishly fed. The queen is capable of laying more than a thousand eggs in a day, and in the springtime she does so. She has no other function.

A *drone* is a male bee that has no other function than to mate with a queen. A queen mates early in her adult life, soon after she has acquired wings and learned to use them, for mating takes place high in the air. She never mates again the several years of her normal life. Therefore very few drones ever actually mate with any queen. After mating the drone immediately dies. When the autumn frosts come and the drones are no longer needed, they are mercilessly thrown from their hive by the workers, to perish in the cold.

All the other bees in the hive are workers. There may be fifty thousand workers in a hive at the peak of the summer. The role of a worker within the unbelievably complex social structure of the colony is determined by its age. Only the older ones fly forth to forage for nectar and pollen. The younger ones remain behind performing janitorial functions, tending the young, building combs and performing the numerous other exact and specialized tasks of the colony. During the active season a worker lives only about six weeks, spending the last two or three of these going to and from the hive and fields, sometimes over distances of miles.

A *swarm* is a large, often enormous, assemblage of bees that has left the hive in order to establish a new colony someplace, sometimes miles away. After issuing from the hive in a roaring cloud, normally during the swarming season of late spring, the bees form a large cluster on any nearby object and remain there, occasionally for several days, while scout bees search the countryside for a suitable place to establish their new colony. This is usually a hollow tree, which will henceforth be a *bee tree*. The swarm consists of the queen mother, a few hundred drones and thousands of workers. Left behind in the hive from which it issued are still thousands of bees and a young *virgin queen*, or sometimes many special *queen cells* from which virgin queens will shortly "hatch," or emerge. The first to emerge immediately destroys all the others in their cradle cells before they can come out. If, however, two such queens emerge and encounter each other they engage in combat and only one survives, this being the only time in her long life that the queen draws her stinger.

I take the honey from my hives by removing the supers and taking them home, having first trapped the bees out of them by a remarkably handy little device called a *bee escape*. The frames of honeycomb, filled with honey and neatly capped over, are then spun in my *extractor* or *spinner*, which whirls the honey out by centrifugal force after the surface cappings of the combs have been sliced off with a warm knife. Other than the straining-out of bits of beeswax, honey requires no processing, although most commercial honey has also beem warmed and filtered in order, as its processors imagine, to make it more "attractive" to purchasers and to retard the tendency of virtually all honeys to granulate or turn crystalline.

Comb honey, once common but no longer so, is simply pure honey comb constructed by the bees within the very containers in which it is eventually sold to consumers. It is one of the most beautiful products of nature, the only strictly natural honey there is and, indeed, the only truly natural sweet known to mankind. It is expensive, because it requires special skill to get comb honey. The relatively few apiarists who have mastered and still cultivate this art are justly proud, for they have forsaken the greater profits of extracted honey for the deeper satisfaction of meeting a more demanding challenge.

As in all crafts, there is a more specialized vocabulary covering the more recondite aspects of apiculture, but these are the basic words and, indeed, the basic facts of this ancient art.

THREE

The Bee Yard

One needs a sanctum, a quiet retreat, a place safe from intrusions, a place to work out one's thoughts, dwell in feeling on the moment, on the sweetness of the hour, the arising and passing of things. This may have been easier when people lived in cabins and got around by foot or by horse, so that a greater part of life was lived at a quiet pace. Perhaps then the solitary hours, at least for country people, were not so hard to find. Now, with engines ubiquitous and the general race against time that seems to characterize the life of most, the hours of seclusion must be sought and are not always easily found.

The bee yard, for its master or mistress, is a natural sanctuary whose visitors are welcome ones. People almost never intrude in my yards, for hardly anyone knows where they are, and the fear of stings forms an invisible fence to any who do. Occasionally, a beekeeper friend may call, or the bee inspector, who is always welcome. But I rarely encounter the bee inspector in any of my yards. He comes only about once a year, and I can always sense his recent presence—the brick of one of my hives is out of place or a burnt match lies where I did not drop it or a tire track tells the tale. I then find his note, saying he was there, what irregularities he may have noticed that I should know about and otherwise declaring all to be well. When I occasionally come upon him, I do not regard him as an intruder. He can talk about bees while being paid to do his work and I can ask all sorts of questions about what he has found on his trips around the countryside, how the honey flow seems to be shaping up and things of this sort. There is a feeling of sodality between us, a sense of common understanding. We possess a shared initiation into the mysteries, and numberless things are instantly understood without needing to be said.

But the bee yard, when not the scene of herculean labors, as at harvest time, is largely a place of quiet where one feels not alone but rather an integral part of the scheme of things. Solitude is not really the word for it. Communion is. One is not separated from company but only from distraction. One's thoughts and feelings are not imposed from without but elicited from within, rising in absorption with the vast surrounding nature.

The hum of the bees overhead, which in spring and during a honey flow approaches a roar, is to me what the sound of the surf is to the beachcomber. It is not a menace or warning, but a reassurance, almost a voice speaking. It would instantly carry the thoughts of others, the uninitiated, to the association with stings. The sight of the bee master, placidly standing in the midst of this roar, would give an outsider no reassurance at all. The rare intruder who comes upon me in one of my yards therefore retreats, and the yard and its master are again as secure as if surrounded by a high wall.

Smaller visitors, feathered and furred, come and go at will, of course, as oblivious to the bees as the bees are to them. The chatter of the birds is unabated, and my appearance produces a squeak from an occasional chipmunk. Off in the meadow a pheasant gives warning to her chicks. But in general all these

living things share the peace with me, and I shall always keep it with them. The bees themselves have very few enemies, and I am glad to move about my yard with the understanding that, from the standpoint of nature, this domain is primarily theirs.

If it is about midday and I have come to work with my bees, I can first rest in the shade with a bit of fruit, my jug of water and a sandwich, my good dog joining me there for the hundredth time. There I get in mind the best way to proceed with whatever needs to be done. I do not know how one could be richer than I am at such moments. Certainly, one's world is not really enlarged with possessions, except by first reducing the world in one's own conception. The setting I am immersed in—the infinite sky overhead, the warm earth, the living woodland, the fields and the many creatures nature never endowed with a sense of property—all this is enough, and the price of it is simply acceptance.

Nothing is despoiled, nothing destroyed or reduced to ugliness or ruin by the presence of a bee yard. The bees themselves forage over literally thousands of the surrounding acres, inconspicuously flying for miles and then returning unerringly to their own hives and no others, but the treasure they bring back is no fruit of plunder. On the contrary, nature has been enriched by their work, for the millions of flowers that have yielded this treasure have in turn received a great gift from the bees, the very gift of life for their seeds. Without the visitations of bees most of these flowers would exist to no purpose, dry up and die without leaving seedlings behind to carry their species into another season.

Honeybees in particular are their saving angels, for unlike other bees and insects that transport the precious, life-giving pollen, the honeybees concentrate upon single species in their foraging. They do not wander aimlessly from one kind of flower to another, which would accomplish little from the plants' point of view. Having started with a particular kind of plant—with an apple blossom, for example—they stay with it. Thus nature and, incidentally, the harvests of men are made more bountiful, and at no price, for the bees also have their reward. One of my yards is tucked off inconspicuously in the corner of the holdings of a large orchardist, who has not only hundreds of acres of fruit trees, but melons and other fruit as well. He knows, as I do, that the honey my bees gather there is a small compensation for the enormously increased yields of his acres.

The beehives themselves involve no assault upon nature or upon the senses. They blend with everything as nicely as a bee tree would. They are, in fact, not substantially different from bee trees. Things have been arranged for the greater convenience of the beekeeper, but the bees take no notice of this. There is nothing unsightly about an apiary, nothing to suggest disharmony. Nothing conveys the impression that, for example, a hydroelectric plant set next to a beautiful stream does or a factory rising in what was recently farmland. On the contrary, an apiary is a lovely thing to see, and beehives are even considered ornamental in gardens and lawns.

The hives need no painting, although there is no harm in doing it if their owner wants to please his own eye. The bees find their way to their own hives more easily if the hives do not all look alike. I rarely paint mine, and as a result no two are quite alike. Most have the appearance of many years of use and many seasons of exposure to the elements. Some were once white, some green, brown, a few even red, having once been painted with leftover barn paint. Most are now weathered, faded and peeling. It does not matter. If I get a hive or super that has never had a coat of paint, which is seldom, then I give it a coat of creosote. This violates the most dogmatic recommendations of the beekeeping manuals: they all contend that creosote never dries and that the bees detest it, both of which statements are false. Before long my creosoted hives have a weathered, inconspicuous and practical look that is pleasing to me and obviously unobjectionable to the bees.

About the only thing that is uniform about my yards, other than the size of the hives, which are all standard ten-frame ones, is that the hives are raised from the ground and arranged in pairs. I find them easy to keep track of that way and I can learn their individual peculiarities. I form a positive fondness for some of them, as though they were friends. A hive stand to support two hives is cut from a ten-foot length of two-by-four, made to form a rectangle and held off the ground by a rock or brick at each corner. Beyond the few cents for the big nails, these stands cost me nothing, because discarded two-by-fours can be found all over for no more than the price of picking them up. It makes no difference if they are filled with nails. They keep the bees up off the damp ground. Dampness is the bees' worst enemy, and the one most easily foiled.

By raising the hives, mice are also discouraged from slipping into them in cold weather when the bees are dormant, and the hives themselves are protected from rot. Of course, many animals find shelter beneath my hives, within the cozy space provided by these simple stands—not only mice, but sometimes snakes, once even a hen and her chicks and once the bees themselves. This last resulted from my hiving a swarm which, incomprehensibly, chose to set up housekeeping *under* the hive rather than *in* it.

Beekeepers are always tempted to build their hive stands with a little ramp sloping down from the hive entrance to the ground. They think that the bees, falling short of their door, can walk up this ramp, but there is nothing to this whatsoever. None of my stands has such an alighting ramp. A bee rarely misses its door. When it does it does not crawl to it. Rather it takes wing and makes a fresh approach, flying right in the second time. The sloping ramps beekeepers feel obliged to supply express some felt need of their own, not of the bees.

Every hive has a brick on top. This custom, quite universal among apiarists, probably originated with the desire to keep the cover on in a strong wind. Doubtless it serves this purpose, at least in winter, and a hive would seem incomplete without the brick. But besides this the brick serves as an excellent device for keeping records, and I can tell at a single glance at a yard any special conditions of any of the hives. The brick in its usual position, fore and aft, means that all is apparently normal with that colony. If standing on end it tells me that the colony was filling its combs with honey last time around and is now either in need of another super or of harvesting the ones that are there. If laid crosswise, the brick indicates that the colony is seriously abnormal—that is has no queen, for example, or that it has laying workers, which is a fatal aberration and will require prompt attention before the colony perishes and the hive is taken over by wax moths. If the brick is lying crosswise on the edge, this means that there is some question as to whether all is well and that I should now have another look.

Usually, of course, all or nearly all of the bricks are in their normal position except during a honey flow or just before harvest. Then nearly all are cheerfully pointing upright, conveying instantly the anticipation of a bountiful honey harvest. It is a rather good system and gives me the additional assurance that no one has tampered with the hives since my last visit, even though little assurance of that kind is needed. If the bricks appear undisturbed I feel that no one else has been near them, for they speak a language invented by me.

Few things are so satisfying to a beekeeper as the sight of a well-tended apiary, especially when it is ringing with the hum of the bees and the hives are towering with supers, presumably loaded with honey. A beekeeper, finding himself in unfamiliar territory, has his eye peeled for the sight of an apiary. I do not know why. Nothing is learned just from seeing an apiary from the car as one drives along, but it is always rewarding somehow. Perhaps it is a reminder that there is still another person who shares some of one's own feelings and joys.

Among beekeepers there is a tradition that they not own, or even in any formal sense rent, the land on which their apiaries are set up. I know of no other branch of husbandry of which this is true. Certainly I would never approach a total stranger and ask if I could put sheep on his land. It is perfectly acceptable to ask that question about beehives. The hives take up little room and, with a bit of inquiring around, one can always find out-of-the-way places for them. It is surprising how many farmers and other country people are pleased, sometimes even eager, to have bees around. Thus I acquire a spot for my bees, for which I do not even have to pay taxes, and the bees themselves acquire thousands of acres of forage land over which they can come and go as they please, gathering the bounty it offers. They do not care who owns those acres, nor do the flowers that bloom in them, nor do I.

Of course I always reciprocate, giving the owner of the patch whatever honey he needs. My rule of thumb is a gift of two pounds of honey for each hive I have set on his land, which may be as many as twenty or so. It is a rough rule, never mentioned to the owner himself, and casually acted upon. Sometimes it is more, sometimes less, but in any case, it is cheap enough. The hive from which he gets two pounds yields perhaps a hundred for me. One owner prefers an occasional bottle of whiskey, left near one of my hives where he, but not his wife, will find it, and this works out all right. I have in the back of my mind a rough idea of when another bottle is due, but we never discuss it. I rarely see the landowners, except when I pull up to their doors with honey for them, and then of course I am welcomed.

How do I find these apiary locations? By asking around. Sometimes I simply go to the door and put the question to whoever answers: How would they feel about having some bees there? Usually the answer is negative, but it is also usually polite. Even those who are appalled at the thought of bees on the place can often point out some crazy neighbor who they think would probably be delighted at the idea. So in one way or another apiary sites are turned up. Of course, as the manuals say, the apiary should be on the brow of a hill that slopes off to the south or the east, but not one of my yards meets those ideal standards. I am content if the spot is sufficiently elevated to avoid dampness, reasonably safe from vandals and convenient to an accessible drive or lane. This last is the one essential require-ment. One can walk a distance to and from his bee yard with pleasure until harvest time. When that time comes one must be able to drive a vehicle close to the hives, for the alternative would be to break one's back with the labor of harvest.

My yards present a great variety. There is none I do not enjoy, but some are more rewarding than others. Some I have arranged with better sense than others, so that my chores there are easier—the hive stands are well spaced, for example, or the wild raspberries and cockleburrs are not a problem. Some reward me with honey flows that come faster and heavier than in others. Approaching one of these yards, I can always hope to find a large crop of honey gathered since the last visit. The hope is often rewarded and thus kept alive from one year to another. Another yard is the occasional source of some unusual honey, perhaps of a kind that is much sought after, such as buckwheat. Here, although my hopes are rewarded less often—sometimes three years or more go by with only disappointment—when the reward comes, it is a foretaste of heaven.

I have one yard in the corner of a woodlot that differs from all the others in that way. Overall, my honey harvest there is less than from the others. The area is agriculturally poor, and one finds there more of the works of nature than the works of men. Farms are for the most part in a state of collapse and neglect. Yet, because the soil is poor, the few remaining farmers do sometimes grow buckwheat. These buckwheat fields are discovered by the bees, but almost never found by me, however long I might search. And sometimes, from the eighteen hives I keep there, I harvest a huge crop of this incredible honey. Some years there is little, if any, although I generally get a decent crop from other plants at this place.

I shall never abandon that spot if I can help it. Apart from the occasional buckwheat, there is no more pleasant place to be, just because it is so isolated, so abandoned by people and repossessed by nature. The corner of the woodlot, which I of course do not own, and whose owner I have seen only once in the past decade, is a tiny Garden of Eden. Because the hives are shaded, my bees rarely swarm there. For the same reason they are never encumbered or obstructed by weeds. The animal population is large and varied, providing a constant source of delight. And sometimes there is that sweet reward that is the ostensible point of it all—honey supers crammed to the last cell with dark, rich buckwheat honey. I need then only post a sign at my cottage door, BUCKWHEAT, and those who know what it means beat a path to the door to get it.

Each yard has its own rewards. The home apiary, which is the only one standing on land that I own, is for comb honey, which makes it unique and exciting. Another is a reliable source of alfalfa honey, while still another is located near wild raspberries. By speading them out over the countryside, I of course increase the distances I must travel to tend them all, but it is worth it, not only for the variety, but also in terms of spreading my risk. In any given year one yard or another may not do particularly well, but it would be rare indeed for all to do badly; in fact, this has never happened. Discouragement in one place is compensated by bounty in another, and I feel more in tune with things when I dwell more on the latter than the former.

FOUR

Spring

Like the bees themselves I come to life in the spring, and my rebirth is as sudden as the season's. Last week the bees were clustered in their hives, miraculously maintaining the hive temperature at 65°F., while winter, yielding reluctantly, piled the bee yard with snow drifts and hurled its winds against the hives. The bees knew it was the season's dying effort, for weeks ago they began preparing for warmth. The queens began laying eggs again before there was the slightest hint of spring to my limited perception of things. They began tentatively, with just a few eggs in the deepest center of the hive, as far from the hostile elements as possible and protected by living blankets of workers, layer upon layer.

Then, as the days gradually lengthened, the brood nest was carried outward.

Thus the bees gained an advantage. Already new members of the colony were at hand, filling the places left by their exhausted predecessors. It is no wonder that the hives, so lethargic when I "put them to bed" in October, burst into life with such explosive suddenness the moment the warm spring air sends the snow into streams in thousands of rivulets and stirs the life underneath the earth.

The colony has been preparing for this day, as I have, they in the darkness of their hive and I in my warm shop, getting ready, waiting, keeping an eye on the portents of spring and the first stirrings of the hives, watching the course of the sun and the changing colors of the willows. What was a desert of snow quickly becomes spangled with the glow of dandelions whose pollen will feed the larval bees emerging in such countless numbers in the hives. Spring peepers suddenly fill the night with their low whistling, reviving in me the sweetest memories of youth. Soon afterward the skeletal black locusts will be heavy with their creamy blossoms, saturating the air with their scent and, if the weather is kind, filling the honeycombs with one of the earliest and loveliest of honeys.

Spring does not creep up on us slowly in these parts, which everyone refers to as "upstate." It and the seemingly invincible New York winter engage in a death struggle, winter reasserting itself time after time, refusing to let go until that moment when resistance becomes futile and spring triumphs in a blaze of glory. Then, virtually overnight, the white of the wild cherry replaces the winter cape, and the life of the beekeeper assumes a new tempo.

Woe to the beekeeper who has not followed the example of his bees by keeping in tune with imperceptibly changing nature, having his equipment at hand the day before it is going to be needed rather than the day after. Bees do not put things off until the season is upon them. They would not survive that season if they did, so they anticipate. The beekeeper who is out of step will sacrifice serenity for anxious last-minute preparation, and that crop of honey will not materialize. Nature does not wait.

My bee yards are scattered about over a wide area, and this

apparent inconvenience allows me to savor spring in its different aspects. My two most distant yards are about seventy-five miles apart, with others strung between. It adds to the driving, but my bees are mostly near routes I must travel from time to time anyway, so my bee trips are often combined with other missions.

Some of these yards are at the edges of meadows, in the full sun. They say that locating them this way gets the bees up earlier in the morning, especially if their hive entrances face east, but I do not entirely believe this. The bees in my woodlot yards seem to do just as well, and they swarm less. Still, some variety is pleasant.

Besides this, I have friends in all these different spots. Out in my meadow yards first thing in the spring I have the company of an occasional pheasant. She lets go with an alarmed squawk to her chicks suddenly warning the whole meadow of my intrusion, even though I have been there for more than an hour. Here and there a snake has sought protection under a hive, with not the slightest objection from the bees. The field mice come too, and one of these can be terribly destructive if allowed inside a hive. The bees cannot deal with him after the frost has driven them into cluster and dormancy in the fall, so the mouse, warm blooded, can ravage to his heart's content. Still, I will not destroy him. It is my own fault for not inserting the entrance guards in time. I am entitled to a few epithets and curses, but the field mouse is still entitled to his life. The beekeeper's craft is made better if he sees it, not as an assault upon nature, but as a pursuit that fits in with the total scheme. The mouse is part of that scheme, as are the snakes, the pheasants, the bees, and indeed, as I am.

In my woodlot yards the scene is rather different from the

meadows. Here the chipmunk is my constant companion, though he cautiously keep his distance. He has no use for bees and never ventures toward them, but near my cottage he makes his winter's bed in the old beehives piled in back of my shed and usually steals a fair amount of my smoker fuel for his bedding. I cannot think of a better use for it, and it cost me nothing anyway—old burlap bags, wads of baler twine and some chunks of rope. The chipmunk sleeps more profoundly through the winter than the bear, I am told, and probably has a regular apartment down there in those hives, with pantry, lavatory and everything. Sometimes he slips into my kitchen when I am away, stealing every last kernel of my popcorn without leaving a trace—which is how I know it was not a mouse. The chipmunk is welcome to that too. Fifty cents worth of popcorn will carry him through a good part of the winter, and I feel I have more than my money's worth.

One of the joys of a woodlot yard is to look skyward in the spring, through a break in the foliage, to see the thousands of bees cascading in like a waterfall and rising in equal numbers to scatter over the countryside for miles around. How they do this without constant collision I cannot imagine. They stream upward and downward without any interference whatever, threading their individual irregular paths with such speed that it would be difficult to follow them if there were not such numbers. They are oblivious to me, even though I may be standing directly beneath the break in the foliage that is their entrance to the yard. They swoop past me on every side, then each to its own hive, which is indelibly fixed in its memory from among the twenty or more hives that are there.

The spectacle is greatest in spring, when each bee seems to feel that the destiny of the race depends upon its wings. I may stand directly in front of a hive, into which the bees were pouring a moment ago until I obstructed the approach, but they do not dream of attacking. They are driven by the need to forage, gather and bring home. But when I step aside, restoring the familiar sight of the hive, they descend upon it in a cloud. They display every color of the rainbow, for they are carrying large and colorful pellets of pollen on their hind legs, gathered from dandelions, willows, rockets and other spring flowers. This pollen is the "bee bread," as beekeepers call it, intended, not for themselves, but for their rapidly developing larval brood. It is an inspiring spectacle. I have kept bees in the heart of a large city, on the flat roof of my garage, where they were sheltered by shade trees and used such an overhead opening in the foliage as their gateway to the world. Although thousands of bees constantly rose and descended through this opening, neighbors only thirty or forty feet away were aware only of the low droning hum of their wings, and sometimes would not hear even this without listening for it.

The first jobs in spring are to make sure that the hives are still clean and that the entrances are not obstructed by dead bees, then to check for the invasion of any mice, to see whether any colonies have perished and which ones may have squeaked through on a prayer and now urgently need to be fed to survive the season's final trial. Those few hives that have suffered severely must be dismantled and their entrances cleared.

In the fall I tilt each hive foreward just slightly, slipping a stick or pebble underneath it to hold it in that position. As a result the melting snow does not run into them and most of the dead bees that drop at the entrance do not accumulate there. My hives are never wrapped or insulated against the winter cold, for I learned long ago that it is moisture, not cold, that works hardship. Each hive is provided with ventilation to eliminate excess moisture. Over the years many have developed cracks and holes, which, if not too large, serve perfectly to insure the desired ventilation. In addition, the bees use these openings for coming and going, for back or side entrances. This introduces a bit of irregularity into the apiary but it somehow suits my sense of how things should be. On newer hives, which so far lack such imperfections, I leave the crown board open to provide ventilation. This permits moisture to escape but keeps wind and melting snow out.

It is a simple matter to see whether the hives are clean inside, for any refuse there is accumulates at the bottom. I insert a long, slender stick into each entrance and remove it with a sweep. If all is well then nothing of significance is swept out. It is pleasant indeed to go down an entire row of hives with this result, for I then know that I have wintered my bees well. Instead of spending the next hour or so dismantling and cleaning bottom boards I can sit down nearby with a banana and my thermos of hot tea, watch the bees and listen to their dense hum, congratulate myself on how well I have done here and in my imagination raise up the crop of honey that is sure to follow.

Some beekeepers, trusting the ways of bees less than I do, at this point routinely "switch hive bodies," that is, switch the positions of the two stories of each hive, thinking that this will induce the queen to increase her egg laying and distribute it more widely through the hive. I doubt, however, that any such result is accomplished, and in any case I have long since found that such planning is best left to the bees.

There is no more doleful sight in a bee yard than a winter-killed colony. I have never lost a colony to dampness since I learned of the importance of hive ventilation, but sometimes a colony is lost to starvation. This is a hard thing to admit, for in principle there is no reason for any colony to perish from this cause. It is simply a matter of my hefting each colony from behind in the fall to make sure it is tolerably heavy and hence amply provided with honey for the impending winter. But what is a simple matter to describe is not always so easy to practice. October is likely to find me rushing against time to get the last of the crop harvested before snow is on the ground and before the late fall honey from the goldenrods and asters has begun to granulate in the combs. Just as spring can come suddenly, so can winter, and one must do first things first. More than one hive is likely to begin the long trial of winter with its meager stores of honey supplemented only by a prayer on the lips of its owner. If everything were as it should be, if I never let myself fall behind and the weather steered a predictable course, if people made no pressing demands upon me at critical times when my bees demand my time just as urgently, then colonies would perhaps never be lost to the winter.

But things are never perfect. I see my bees for the last time in October. By then the first frost may have come, the drones have been hurled from the hives and the bees are languorous even on the occasional warm days. The flowers are all gone and the gray skies promise a sudden snowfall at any moment. It is a rush just to get the entrance wedges all in place to keep the mice out. The bees are now resigned to the long and difficult months ahead. Some of the hives are nice and heavy. I have no idea what they weigh, but experience has taught me that when they feel that way all is well. Others I wish were a bit heavier but it is past doing anything about it now. The honey they need has gone to the honey house instead, and some of the fall flowers I counted on failed. I shall worry about these hives, but when spring returns, it is always surprising to see how well they made it. A few may have succumbed, if the winter was long and hard, and many more may have made it only by a hairsbreadth. With the very first appearance of any bloom these will be entirely rejuvenated.

One year, when the goldenrod in the area failed that fall, a whole yard was light. I thought I would be lucky if three of the sixteen colonies survived. Thirteen made it, and of the three that did not, two had been toppled by vandals. It was a gratifying reassurance to check that yard in the early spring, with snow still on the ground, to peer ever so briefly into the crown board opening of each hive and to see that the bees were alive. By the time the dandelions had come and gone, each of those hives was again as heavy as stone, loaded with the treasures of honey and pollen.

It is perfectly easy to tell whether a colony has perished from starvation rather than some other cause: there is no honey in the combs. What one finds instead are dead bees, masses and masses of them, tightly clustered, the combs filled with them, so that each cell has become a tomb rather than the larder it was meant to be. The dismal spectacle appears hopeless—moisture has invaded the hive and turned its contents to a "mouldering mass of sodden corruption," in the perfectly chosen words of some writer of long ago. Vermin have moved in, parts of the combs themselves disintegrate at touch and the temptation is to cast the whole forsaken mess upon a good fire.

Textbooks in beekeeping rarely mention this problem. Evidently their authors are reluctant to admit that it arises. The beginner is thus left in the dark, depressed and dismayed. There is no need for this, however, for here one learns again of the bees' capacity to work together in the performance of unbelievable feats.

I dismantle the dead hive and scrape its bottom board, which takes only a second or two. Next the two stories of the hive are separated and the combs from each removed, one at a time. Most of the dead, soggy mass of bees falls to the ground as combs are separated. The rest I scrape loose with my hive tool, but there is nothing I can do about the thousands of bees filling the cells of the comb or about the bluish mold that has settled. No matter. I do what I can and replace the combs. Now at least air can get inside, the mold is checked and the hive will begin to dry out.

In a week or two, when the opportunity presents itself, I hive a swarm in this salvaged hive. Since it is spring and the season of swarming there is usually no dearth of swarms for this purpose. From then on it is as though nothing had happened. The bees clean the hive and combs within a few days, repair broken comb and restore everything to its original condition. The truly monumental work of apiculture is always done by the bees themselves.

Sometimes a swarm takes over a dead hive at its own initiative, greatly simplifying my work. It is a great satisfaction when I find this has happened. I go to my yard, set for the depressing task of cleaning up a dead hive and discover that the hive is quite alive, that it has been resurrected by an industrious swarm, and that thousands upon thousands of bees are doing the very work I had planned for myself. One winter, because of an early frost that destroyed the fall flowers, my losses were very heavy. My best yard, normally stocked with twenty-two colonies, lost five, which is unusually severe. I was late getting these five cleaned, and by the time I finally did, three had been taken over by swarms. I gave them a hand, cleaning out the bulk of the dead bees and decay and clearing the bottom boards, but it was astonishing how much these bees themselves had already done. The swarms had not been there long, for the queens had only recently begun laying eggs.

On my early trips to the yards in the spring I take jugs of sugar syrup to carry over any colonies that may be in need of emergency relief. No hives are opened yet, for there may still be patches of snow on the ground. I simply heft them from behind, an inch or so, to get the feel of their weight. Any light colony is given a gallon jug of life-saving syrup. It tips the balance. I pick up these wide-mouthed gallon jugs at lunchrooms, usually at no cost, where they were used for mayonnaise, relish and such. I punch four small nail holes in the center of the lid, dump in a five-pound bag of sugar, fill the jug with hot water and stir. The jug is them inverted over the crown board opening of a light colony and the bees empty it in a week or less. Of course there is no danger of any of this sugar syrup finding its way into honey, for the bees consume it at once, and in any case no extracting supers will go on the hives for some time.

Spring is the time of gladness for a beekeeper, not because of what he has accomplished, of course, but for what is promised. This is when I am nourished by hope. The season lies ahead. I always feel sure it will be the record one, and past disappointments do nothing to dampen that hope. Life quickens on every side. The signs of winter's decay subside and are replaced by renewal, not only in the hives, but in the warm earth, in the meadows and woods, and in the air, where every motion and sound declares life. Now is when I not merely know what I have known all along, but deeply feel what I know, that I am alive and intensely aware, in every cell and every thought and feeling, and I rejoice.

FIVE

Swarms

A beekeeper has few memories that rival those of the swarms he has seen. It is a thrill unique to this craft. It seems to me I remember all I have ever dealt with, hundreds by now. My thoughts still go back to a distant boyhood, and I see myself bicycling home, barefooted, a large swarm of bees enclosed in a burlap bag and held at arm's length so that none would be injured as I peddled along. The thrill and fascination that filled me then as I watched large swarms stream into hives has never weakened. I still sit in silent wonder, when now, alone, in one of my yards, I hive a new swarm. It follows exactly the pattern established millions of years ago. It is at this point in the colony's cycle that the psychology of the bees becomes most wondrous. It is now that the habits and orientations that have governed their every movement from the moment they first took wing are abruptly halted, to be replaced by a totally new orientation, as though the former had never existed. It is as though they were determined by a destiny laid down in advance. The phenomenon draws a philosopher's attention to the mystery that inhabits every pocket of creation. We see only a small part of the surface of things. The rest will be forever hidden from us, to be appreciated for its felt but unfathomed presence.

Swarms almost invariably issue before noon and, in my latitude, in the months of May and June, reaching a peak of frequency early in June. By the Fourth of July I have ceased giving them much thought. If several days of rain in the spring are followed by clear, warm mornings, then swarms are likely to issue from the hives in great numbers. Of course wet weather does not by itself cause swarming. The underlying causes are much deeper and are not really understood by anyone.

The preparation for swarming takes about ten days, eight at least, for bees will not swarm until they have reared a number of new queens so that one of them can remain behind to perpetuate the parent colony. It is seldom that uninterrupted rain lasts that long. Therefore, the colonies that swarm in such numbers following wet weather have been preparing to do so before the rain began; the weather only forced a postponement. Although it is considered one of the few near certainties of apiculture that bees will swarm on a warm day that follows days of rain during the swarming season, I have nevertheless made the rounds of my yards on such a day without finding a single swarm. On other days, when for one reason or another I expected to find few or no swarms, I have found a dozen.

Hives headed by queens three or more years old are almost certain to swarm. A colony with a queen less than a year old will seldom swarm unless goaded to it by an inept apiarist. This suggests that swarming might be controlled by requeening one's colonies every year, and in fact this is good advice to one who has only a dozen or so hives. One who has a hundred or more must be content with a less perfect scheme, for otherwise one's beekeeping would consist of little more than requeening.

It is not clear why old queens should be the ones to head swarms, and yet there is a certain abstract intelligibility to it. Although old, she is perfectly capable of founding a new colony with her worker daughters. When she is gone, perhaps soon, one of these daughters will have been miraculously transformed into a new queen, assuring the ongoing life of the colony. In the meantime a new young queen has remained behind to perpetuate the "parent" colony.

Given the spectacular character of swarms of bees it is surprising that they are so seldom seen. Even beekeepers, who are constantly on the alert during the swarming period, seldom see them except in their own bee yards or when summoned by someone else who has discovered one, probably for the first time in his life. The swarm, after it issues from the hive, is in the air rather briefly. For those few minutes it is awesome. The sight fills the beekeeper with excitement and, if the swarm is from one of his own hives, with consternation, for he sees in it the destruction of his hopes for a large harvest of honey from that hive.

People unfamiliar with bees and unattuned to nature are filled with terror by the sight and flee with their hands on their heads. In fact, there is nothing personally threatening about a swarm of bees. Without being hurt one can stand in the very midst of it, the thousands upon thousands of bees swirling about on every side, a few alighting on one's head or on an ear or a shoulder, then taking wing again to rejoin the foray. Provided one gets one's spirit in tune with the excitement and is careful not to crush a bee inadvertently or to let one become entangled in clothing or hair, the chance of being stung is exceedingly remote.

After such a commotion silence soon falls upon the scene, and the swarm now hangs quietly and inconspicuously on some nearby branch, bush or post. The colony that produced it has resumed its accustomed activity, as though nothing whatever had happened, and even the most expert beekeeper finds it hard to say which colony this is unless he was there to see the swarm come out. Now one understands why swarms are seen so seldom. It would be easy to stand within a few feet of this clustered mass, composed of perhaps twenty thousand bees, and not know it is there. Even an experienced beekeeper seldom finds stray swarms. In his own bee yard it often happens that a swarm, which he had not suspected was there, suddenly breaks cluster almost at his elbow, and he stands in helpless dismay as it moves off across the meadow in a vast cloud, never to return, to store in some hollow tree all the honey that he thought would be his. Now is the time for the beekeeper to remind himself that nature recognizes no principle of human ownership. Certainly the bees do not.

During the swarming season people discover clustered swarms in their yards in what seem to be extraordinary places. This is when I receive telephone calls from persons I have never heard of, whose anxious inquiries have turned up my name, as well as from the police and fire departments. The bees' choice of a clustering point is generally arbitrary. It can as well be some lamp post or traffic light as an isolated tree in the country. I have been summoned to the heart of the city to deal with a swarm clustered on a fire hydrant and arrived to find the street barricaded and the police directing traffic clear of the danger point, keeping the curious at a safe distance and being careful to maintain that distance themselves. This is evidently one of the hazards of police work not dealt with in police manuals. To appear in such a setting clad in sandals and tennis shorts and

deal casually and expertly with the cause of all this alarm, accompanied by many "ohs" and "ahs" from the bystanders, is a pleasure that can be matched in few crafts.

The occasion also offers a good opportunity to convey to people that the danger exists only in their confused perception. Bees possess stings only for the protection of the hive. Swarming bees have no hive to protect, and therefore no inclination to sting. I gently dislodge them into my big swarm funnel and swarm box, occasionally brushing a stray bee from my ear or nose or clothing, then drive off, as tranquility settles once again on the human scene.

People are sometimes terrified to discover a swarm in their yard and will eagerly pay an exterminator any fee to dispose of it. If boys discover it they treat it as a target for stones. Often the impulse to mutilate and destroy seems to arise spontaneously in people when they are confronted with something unfamiliar in nature. More than once I have been told of a swarm of bees and, arriving upon the scene, have found ashes, char, and the odor of kerosene, and many thousands of lovely honeybees scattered on the ground, dying or dead. One would think it enough to tell people that bees are essential to the natural ecology and agriculture. Those who cannot understand this will hardly understand that nature labored millions of years to perfect these beautifully complex creatures, and that they have as much right to be in the world as their frightened tormentors.

The swarming season is always filled with surprises. One of the happier ones was the most recent. I saw a huge swarm in the very top of a large apple tree near my home apiary, far beyond reach of any combination of ladders and poles that I could assemble. So I ruefully abandoned it to nature. Within an hour, standing again near that spot, I was astonished to find this

enormous limb on the ground, as though by a divine interven-
tion, and there, before my unbelieving eyes, was my great
swarm, confused but still together, within a foot of the ground
and a few feet from the empty hive into which I forthwith
dislodged it. That limb had withstood storms and gales for the
better part of a century, and then chosen this moment, when
there was no wind, to collapse. The weight of the swarm seemed
hardly sufficient to make the difference.

Driving to and from my yards one spring, a spring when
swarms had been unaccountably scarce and I was wishing I
might come upon a few more, I suddenly had the fleeting
feeling that I had espied one from the corner of my eye, a great
distance away across two widely separated lanes of highway and
in the middle of a woodlot. Just for that instant it seemed to me
that I had caught a glimpse of its catenated silhouette against a
patch of blue sky. Such discoveries almost always turn out to be
illusions arising from hope, as the "swarm" materializes into a
collage of leaves or the refuse of tent caterpillars.

This one, however, turned out to be real, as I found when I
had maneuvered my car back to the spot. There were no apiaries
around there that I had ever seen or heard of, but that fact was
not by itself remarkable. Swarms often travel great distances
and recluster along the way. I took the spare hive that I keep in
the car for just this purpose, set it under the swarm, which was
only a few feet from the ground, dislodged the swarm in front of
it as I had done countless times before, then continued on
toward home and supper. A week or so later I returned to find
the hive where I had set it, now taken over by the bees, as I had
fully expected. Since the day was wet and few bees were flying, I
slipped a screen into the entrance and carted the hive off to my
buckwheat yard in the woodlot. That was many years ago, but I

can still point to the colony established that day. Now I can never drive past the spot where the swarm was first seen without looking over to it again, as though expecting to see another. In fact, I like to recall how I discovered it, over that considerable distance and while driving along, for it was truly a feat of sorts, the kind of thing that happens very rarely in a lifetime. I take pride in it, even though it is of no significance to anyone else in the world. That colony has never failed to store up a good crop.

So just as nature sometimes reclaims my bees and distributes them to the safety of tree hollows, so also does she sometimes freely give.

Few things are so remarkable as the abrupt change in the bees' psychology that is wrought by their swarming. Normally, they mark the location of their hive with precision, unerringly returning to it after a foraging flight of many miles. If the hive is moved as little as a foot to one side the bees go first to the spot where it had been before making a fresh approach. This orientation is totally disrupted by swarming, so that the bees appear to suffer a complete loss of memory. If, for example, the swarm is hived, and the new hive set alongside the very hive from which the swarm issued, the bees will nonetheless disregard the original hive—the one into which they were born and to which they had unfailingly returned hundreds of times. Nothing will induce them to enter it again so long as their new hive is there.

Indeed one would suppose that any stray bees from the swarm that were in the air or off scouting when the swarm was hived would easily find their way back to one of the other of these two hives. They would be welcomed at either, although instantly assaulted by the guard bees if they attempted to enter any other hive in the apiary. But they cannot find their way to the hive from which they issued, nor to the one next to it, in and out of which their sisters now freely pour, even though these hives may be only a dozen feet away. Days later one can return and find these stray bees still aimlessly circling, their orientation changelessly fixed to the cluster spot the rest of the swarm has abandoned. They retain their futile attachment to that spot until, days later, they perish within sight of what was so recently their home.

On the other hand, if the queen is removed from a clustered swarm, within only a few minutes all the bees return to their hive—not in a random and haphazard way, but en masse, as if at a signal. So they have not forgotten at all! One sometimes reads that if a handful of bees from a clustered swarm is shaken about in a bag of flour and then released, one can then tell from which hive they came by seeing to which one they return. This has never worked for me, and I believe it is something writers have fabricated. The dusty bees in my experiments were irrevocably oriented to their swarm cluster, and no signal that I was capable of producing was able to budge them from it.

How are such things explained? No one knows. Perhaps someday someone will be able to explain them, perhaps not. But I believe there are bees, indistinguishable from the others, who somehow direct the rest, and that the swarm is entirely dependent upon signals from them, instantly understood by the bees but still hidden to us. One naturally imagines that the queen would play this role, so appropriate to a monarch, but this is certainly false. The queen is the most helpless, least resourceful bee in the entire swarm, displaying nothing but fear, stupidity and an eagerness to do whatever her daughters tell her to do.

Why do bees swarm? One might as well ask why the heart beats, why birds nest and spiders spin, why the earth turns.

Abstractly, the phenomenon is quite understandable. The perpetuation of the species cannot be assured simply by colonies of bees becoming ever more populous. Beyond a certain point the size of a colony serves no purpose whatever. Although a colony of bees in a tree hollow might become ever so populous, this does not in the least insure that the tree will not be lost to a fire or to the ax of a woodsman. From the standpoint of the life of the species, then, nothing is accomplished by sheer numbers. If, on the other hand, that colony has given rise to others like itself in other trees and hollows, then nothing of significance is lost when it perishes itself, as sooner or later it must. Other colonies are now in the world to carry on the life that it is the mission of every living thing to perpetuate.

SIX

Controlling Swarms

While there is no such thing as *the* cause of swarming, there are certain factors that precipitate it. The main one, already mentioned, is the diminishing fecundity of the queen as she ages. Another, also mentioned, is a period of rainfall followed by a burst of sunshine and warmth. Overcrowding within the hive does not, as beginners often suppose, cause swarming, but it does aggravate the tendency. I therefore make the round of my yards early in May to get the first supers on the hives, giving the bees room to store honey and something to distract them from swarming. Any obstructed entrances are also now opened, even though there may still be a frost or even another light snowfall. One shallow super is given to each colony the first time around, then another in a couple of weeks as honey begins to appear in the first ones, then a third a bit later and, in the case of very strong colonies, eventually a fourth.

The colony is thus able to grow and expand, and my eyes are treated to the sight that, from the beekeeper's point of view, is the purpose of it all—new, crystal-clear honey in the combs and snow-white flecks of wax capping it over. The sweet odor of the apiary tells me, before I look, that the bees have found great sources of nectar and that the weather has favored the gathering of it. Colonies that a couple of weeks ago would be lifted almost effortlessly would now strain the back of any strong man; the difference is the honey that is rapidly filling the supers.

Of course I hate to see my colonies swarm, as this means the depletion of the force of foragers at the very moment they are needed for gathering nectar. It is a blow to the spirits to open a hive that a few days ago was boiling over with bees and find its population depleted. Foraging is brought nearly to a standstill just when the fields and meadows and trees are filled with the blossoms of clover or yellow rocket or locust. There is not much change in the apparent population of the hive below, but the supers, where the honey goes, are depleted. The swarmed colony will eventually regather its strength, and with decent luck provide a fair crop of honey from the autumn flowers, but it will be less that I had hoped for.

Swarming is of course essential not only to the survival of the species but also to nature itself, for without bees the many plants—both wild and cultivated—that depend upon them for the viability of their seed would also be threatened with extinction. They even have their place and value to apiculture. A newly hived swarm is usually the hardest working colony in the yard. If hived on foundation in a single-story hive (but not if hived on drawn comb), it can usually be depended upon to make beautiful comb honey because of its excessive industry. Swarms also draw foundation beautifully and can be utilized to get perfect brood combs that will last for decades. These are the very foundation of practical beekeeping. Swarms are also useful for making up for winter losses and getting the hives back to use before mice or wax worms take them over. And finally there is the sheer delight in gathering them, adding needed colonies when one is expanding a beekeeping operation.

I have heard it claimed that by picking up stray swarms, one risks bringing disease to the yards, to which I can only say that none of the countless swarms I have added to my yards over many years has resulted in a diseased colony. A colony weakened from foulbrood, a bacterial disease that attacks honeybee larvae, would not be likely to throw swarms, and in any case the risk of disease is greatly reduced by hiving swarms on foundation. Any diseased honey is thus converted to beeswax and its potential danger is eliminated.

There seems to be no practical way of preventing swarming, and sooner or later beekeepers need to conquer their anxiety over it. One must here, as in so many things, be content with something less than perfection, doing what one can to reduce swarming and, beyond that, hoping for the best. Someone with only a few hives, who does not depend on his bees for any significant part of his livelihood, can undertake herculean preventive measures that sometimes work, but this is no pattern for the beekeeper whose bee yards are spread over the countryside and who must make every hour he spends with them count.

If the queen's wing is clipped she cannot, of course, fly off with a swarm, and there is no harm to her in clipping it provided she has mated. The only time she will need wings is when the swarm issues. Without them she falls helplessly in the grass in front of the hive while the swarm swirls overhead, but they will not leave without her and in moments they return. They won't stay, however, for virgin queens with very good wings are about to emerge within the hive, and in a day or two the swarm is ready to take wing again, this time with one of the virgins. Hence nothing has been gained, unless the beekeeper happened to be on hand to see the original swarm come out. In that case he can put a new hive on the original stand and let the swarm with its old clipped queen run into that, moving the parent colony off to a new spot.

Some beekeepers go back one step further in their effort to combat the swarming impulse, destroying all the queen cells in the hive. This sometimes works. The bees will not swarm without leaving behind cells to provide a young princess to fill

the place abandoned by the queen mother. The bees, however, driven by the swarming impulse, replace the queen cells as fast as the beekeeper mutilates them. Normally this replacement takes at least ten days, but in this war with their supposed master they sooner or later decide to swarm on a seventh day, before the queen cells are even sealed, and the work of trying to keep them eliminated has all been wasted. And what work it is:

examining the frames closely one after another every week, greatly disrupting the colony each time, and then in the end losing the battle as the bees fly merrily off.

Beekeepers have invented all sorts of clever manipulations to combat swarming, all of which involve separating the queen either from the young brood or from the foraging force. Bees will not swarm unless queen, brood and foragers are all present. Thus, for example, the brood is all moved to the top of the hive and the queen confined below on mostly empty combs. There are many variations on this idea, but all involve considerable extra work and equipment, which makes them of doubtful value—and still the bees may swarm.

Probably the only reasonably sure way of preventing a swarm is by what is quaintly called "shook" swarming the colony. To speak of this as prevention is misleading, however, for it really involves the beekeeper's creating a swarm. All he gains is the choice of time. The method consists of setting the entire colony off to one side, putting a new hive in its place, and then shaking perhaps half or more of the bees off the combs in front of the new hive, together with the old queen. The original hive, now on a new stand with all the brood, eventually raises a new queen, or can be given a new one, and both colonies then gather strength and eventually store a fair crop of honey.

The effect of that operation is the same as if the bees had swarmed and the swarm had been hived at its original location but in a new hive. The foraging bees return to their old stand, greatly augmenting the force of bees there, and the moved colony, deprived of its foragers, does not swarm. It is all rather fun, if the beekeeper has time for this sort of thing, but then he does end up with two colonies where before he had only one, and he may not have wanted this. A variation upon this system involves hiving the shook swarm, not in a regular hive, but in a shallow super of foundation and then using it to raise comb honey. This can sometimes be done with spectacular results.

If a colony is moved to a new location in the same yard on a warm day when the bees are foraging, it seldom swarms unless it was on the very brink of throwing a swarm when moved—all the foraging bees return to the original location. But then, of course, the beekeeper has the problem of providing a new hive at that original location. Otherwise, the foragers, returning by the thousands, have no place to go.

Sometimes one can simply interchange a strong colony, which is likely to swarm, with a weak one, which is not. In this case the weak one is strengthened by the foragers of the other, which in turn is not likely to swarm. These same foragers, however, are quite likely to murder the foreign queen that they suddenly find occupying the hive at their location; so what was a promising colony is now likely to dwindle to nothing.

Still another method I have used with some success, but which is really only a variation upon these, is to remove the strong colony to a new location in the yard, and on its old stand put a new hive with three or four combs and the queen from the original hive. This usually works but again one ends up with two colonies where before there was one.

After trying them all I decided some time ago that swarm-prevention measures are a will-o-the-wisp; each has its price and none its guarantee, for bees have ways of confounding the most resourceful of schemes. Bees are not like the domesticated mammals and fowl over which human beings have so totally triumphed. They are still untamed and uncorrupted. I hope they will always remain so. If men were to conquer the bees, bending them entirely to human ends, then there would doubtless be rewards of some sort in terms of increased honey production. But something very precious would be lost. The bees we know care no more for our hives than for a hollow tree; they are as much at home in one as the other, and during the period of swarming they are ready to abandon either for the other. The step to nature is for them a very short one. Indeed, it does not even exist. By inducing the bees to make their homes in the hives we provide, we gain control over their location and enable ourselves to reach into those homes and take what we will, but we gain little control over the bees themselves.

I profoundly hope it will always be so. The claims of human beings are not the only ones that are made in the natural realm, and there is something to be said for the spirit of adaptation and acceptance instead of conquest. We have conquered so much, and in the process laid waste to so much. If now this tiny insect adamantly refuses to yield its own nature to us then a certain balance of things will remain as it should. The beekeeper will always be there, with the hum of the oblivious bees overhead. When he sees a prime swarm fill the air and move off to the woods, following the pattern set millions of years ago, he can draw from the sight an inner satisfaction that transcends consideration of material gain.

SEVEN

Swarm Gathering

A beekeeper can hardly sit still or put his mind to any other work at the peak of the swarming season. Over the years I have managed to bring this anxiety under control to some extent, at least to the point of more or less putting out of my mind what might be happening in my more distant yards when it is not possible to find out. Nothing is gained by imagining large prime swarms there. Experience has taught me that if I rush to my yards, summoned by such specters, I will probably find no swarms anyway. Still, it is not easy to relax on one of those warm spring days when I am tied to other responsibilities. The rest of the world goes on in a state of relaxed enjoyment, as though nothing were different, while a beekeeper feels as though he were sitting on pins.

Sometimes, though, that warm spring breaks on just the day I have set aside for a tour of my yards, and this, if all goes tolerably well, is like heaven. There is a thrill in gathering swarms, especially if I have suffered winter losses and have some hives standing empty. At this time of the year every apiary can be a source of surprises. I might find an enormous swarm, several, none at all, or perhaps three or four tiny ones clustered here and there. These little "after swarms," as they are called, have issued with virgin queens after the prime swarm and mother queen have left. It is not clear why they emerge, or why they sometimes do and sometimes do not. They seem to be symptomatic of the colony's "swarm fever," as beekeepers call it, which is sometimes slow to abate. In any case, they are a delight to find if one is needing swarms, for even one of these, smaller than a football, can grow to a powerful colony and fill a super with honey the first year. Just the virgin queen by herself is worth a considerable sum as soon as she is mated.

The unique reward of any swarm, particularly a large one, is seeing it. Of course, a swarm gathered early in the season represents a potential honey crop of significant value, but a beekeeper who lets his thoughts stop there views his craft with an astigmatism. The manner in which a swarm takes over a new hive is always the same, and the thrill of seeing it is always the same, even though it may be the hundredth time.

I dump a prodigious mass of bees in front of a hive, as casually as if it were a bushel of beans, usually onto a cloth that I have spread to prevent entanglement with grass and weeds. There is a momentary confusion as bees move aimlessly and a few of them take wing. Then a few approach the hive, recognize its potentiality and signal to others by raising their tails high and fanning their wings. The signal is quickly picked up by the other bees, which duplicate it, and within seconds most of them are facing the hive, heads low and tails high in the air as though bowing to Mecca. The mass of bees starts approaching the hive, slowly at first but gathering speed as they come closer and start pouring in.

At this point it is always worth searching for the queen. She can usually be found somewhere among those thousands upon thousands, clumsy and frightened and beautiful to the eye of a beekeeper, and having the least idea of any bee in the throng what is going on and what is expected of her. Certainly she in no sense leads the other bees. If she is removed, however, it produces a visible consternation throughout the entire swarm. The new hive, which roused in the bees such excitement a minute ago, now ceases to have the slightest interest to them until they have her back. When I find her I usually pick her up gently, especially if she seems headed off in the wrong direction, and drop her near the hive entrance. She dashes in at once, away from the unfamiliar daylight and back to the security of darkness. The bees know approximately when this critical event has occurred, when their queen is safe in the hive, and a sense of contentment settles over the entire colony.

This is their home, from that moment and very likely for decades to come. What was before a lifeless thing, without significance, becomes now the foundation of their city and their destiny. From miles around they will henceforth return to this spot, and this one alone. In the course of a few minutes it has become the center of their universe, any other object on earth having meaning for them only in relation to it. They will build it up and protect it with their lives. Nothing will be permitted to befoul this hive, nothing other than a bee will be permitted to enter it, and indeed no worker bee except one belonging to this hive. As these bees themselves perish and are replaced by new generations, they will fall outside the hive or be quickly carried out.

The ease with which a swarm is hived is astonishing, but of course not every swarm gets hived. Sometimes the swarm is high, inviting the beekeeper to risk a broken neck. More likely, he simply does not see it until too late. Sometimes I first learn of the presence of a swarm nearby by hearing it take wing, and sometimes a swarm takes wing right before my eyes just as I am making preparations to gather it and then, as all the literature of beekeeping expresses it, "absconds" over the horizon while I watch in dismay.

It is an inspiring spectacle, if one can overcome frustration sufficiently to appreciate it. The cluster of bees that has hung there sometimes for days, until this moment placid, quiescent,

hardly resembling a living object, suddenly begins to disintegrate, as though it had received a signal. Within a minute the air is filled with a vast cloud of bees hovering at the spot and keeping themselves together as one. The purpose of this is to insure that the queen is present, to make sure she has had time to take wing and join them. The destiny of the colony for generations and, in fact, for eternity rests in the tiny cells of her body, and there alone. Without her, the swarm would be meaningless. Soon the cloud rises and expands, the individual bees threading their way within it in a pattern of zigzags. It begins to drift away, and for a few minutes one can accompany it over the meadow. Then as they become assured that all is as it should be, the bees gather speed and vanish over the hillside, leaving behind the vacuum of silence, as though nothing had happened. In the presence of such a sight a beekeeper can rise above his limited animality and, for a moment, sense, if not understand, the eternal assurances at which nature forever hints.

I have worked up a few simple swarm-gathering tools and techniques, and the tools are always at hand during the swarming season. One of these is a strong ten-foot pole with a heavy hook screwed into its end. This brings most swarms within my reach, though sometimes I must also stand on top of my truck. Another device, and without doubt the best single discovery of my beekeeping career, is an enormous funnel, about two feet wide at the top and narrowing to about four inches at the neck. Rather than hiving swarms where and when they are found —which is strangely a common practice among beekeepers—I merely dislodge them into screened swarm boxes, using the funnel, then hive them at times and places of my own convenience. In addition to these two special swarm-gathering implements, I have the usual ones—pruners, both large and small, a small buck saw, a length of cord and clothesline and a few small queen cages.

Ideally a swarm should select as its clustering point the end of a fairly small branch of worthless sapling about four feet from the ground. And actually they do that, more or less, fairly often. I then merely prune the branch near the cluster, trim away twigs and leaves, and either dislodge the swarm into one of my swarm boxes or carry it to a nearby hive in need of occupants and dislodge it there. If I use a swarm box, I can cart it home and, if necessary, store it in my cool basement for a few days, then deal with it at my leisure. The bees do not mind this in the slightest, just so long as their queen is with them.

Usually, however, swarms are found clustered in less convenient places—in a thornbush, for example, or a tangle of vines. I prune it away as best I can, taking care that the cluster does not get broken up or scattered if this can be helped. If at all possible in such cases it is best to get the funnel and swarm box under the bees, then shake them all in as well as one can. Bees that are left outside will cluster on the outside of the cage in a few minutes, provided, again, that the queen is inside it.

Of course, the swarm is likely to be overhead. The pole will usually bring it down to within reach. If it would require a ladder to get the swarm I usually force myself to ignore it, to pretend I never saw it, and proceed with other things. No swarm is worth the risk of getting hurt.

If, however, I feel I must recover the swarm, then the following strategy usually works. I tie a small stone to a cord and get it over the branch near the swarm, either with a fishing pole (if one is at hand) or an accurate throw. The stone is joggled down to within reach and a clothesline tied to the other end of the cord and pulled on over the branch. Next I tie a small basket to the end of the line and put a comb containing unsealed brood in this. The brood comb can come from any hive. Basket and brood are now raised to the bees so that they touch it. In a half hour or so they become clustered around that comb and in and around the basket, and can be lowered and hived very simply. The last time I used this somewhat troublesome technique I was glad I had, for it was a large, early-season swarm. It made more than a hundred sections of comb honey for me that first season. The time spent on that one was very well spent indeed.

I have oftened wondered how I ever got through season after season of beekeeping without my swarm funnel and swarm boxes and I wonder still why these are not basic tools of other apiarists. They shorten the time spent in apiaries during the swarming season and put me in control of the bees instead of the other way around. Traditionally, beekeepers cart empty hives around to their yards and hive swarms in them as they are discovered. It is not unusual for me to find a half-dozen swarms in a single tour of my yards, and twice that number is not unheard of. Sometimes I will find several hanging about in bushes and trees in a single yard. Without a funnel and swarm boxes one needs to have perhaps a half-dozen hives on hand at all times while checking the yards for swarms.

Nor is this the only problem. Bees once hived do not always stay hived, especially if the operation is performed in the heat of the day. Their cluster point still draws them like a magnet and the beekeeper, after a half hour's struggle to get a swarm down, often faces the frustration of repeating the task or even seeing the swarm he thought he had properly hived take wing and fly off. Again, it is not always practicable to bring a swarm to a hive; sometimes the hive must be taken to the swarm for it is

sometimes impossible to move whatever it happens to be cling-
ing to—a tree trunk, thorn bush, heavy post or something of
that sort. The bees must then be hived at that spot, which is
usually not where the beekeeper wants an additional hive. Once
the bees are hived there that is where they will forever want to
remain, so one must either leave things that way or come back
later and move the hive to a new spot miles away.

All such difficulties are abolished with funnel and swarm
boxes. It usually takes only a few minutes to knock a swarm into
a swarm box, which can then be handled at the beekeeper's
convenience and at whatever yard he chooses. There is no
returning to pick up hives, and the swarm boxes themselves
take up little room in a truck or car.

I also mentioned small queen cages as being among my
swarm-gathering tools. What are these for? Actually, they are
seldom of use, but they take so little room and are sometimes of
such great value that it is worth having them along. I often spot
the queen, and occasionally, if I am quick and alert, I can catch
her in my hand. Once she has been popped into a queen cage
there is no problem. The swarm is then entirely in my control,
for the bees will not abandon her, no matter where she is. If I
slip the queen, cage and all, into a hive with a handful of bees,
then the rest of the bees will soon follow. Or I can suspend her in
her cage with a bit of wire from a convenient small branch and
soon all the bees will be on that branch. Or I could, though I
never have, clip cage and queen under my chin to produce a
large beard of live bees. This is a standard stunt of beekeepers
and it never fails to stun audiences.

Capturing the queen is always fun as well as useful. The sheer
exercise of skill and dexterity in plucking that one bee from the
midst of her thousands upon thousands of daughters is a deep

satisfaction. Once I even plucked her from the air. I heard the swarm emerging from a hive I kept on my porch, and as I stood there watching them I suddenly had the sense of spotting the queen in the cloud of insects that filled the air in front of me. With a quick grab I suddenly, miraculously and unbelievably had her in my hand. The rest of the bees then clustered near the door of a fairly disagreeable neighbor who had always been far from enchanted by my keeping bees there. I slipped the queen into a queen cage and fastened this on a bush in my own yard, next to a black sock that the bees immediately noticed, and within minutes the rest of the bees all came there too, before my neighbor was aware that anything had happened. On other occasions I have spotted the queen at the hive entrance just as she was emerging with a swarm. She is always worth watching for there, if one knows that a swarm is issuing, for she seems never to be among the first to come out. If you are very quick you can pick her from the alighting board before she spreads her wings. I have never been able to pluck the queen from the swarm after it has clustered, but often she can be found as soon as the cluster is dislodged. Occasionally, the bees themselves betray her location by forming a tiny cluster around her.

When a swarm is awkwardly located and it is impossible to get the whole thing in one stroke there is always the question of whether one has the queen. Is she in the swarm cage or still with those bees at the cluster point? Or is she in any of several possible spots to which the bees have been scattered? This question is always easy to answer by noting the behavior of the bees themselves. If the bees in cluster settle down contentedly the queen is with them. If they are agitated she is not with them. If large numbers of bees are bowing in more or less the same direction and fanning their wings, perhaps over an area of several square feet, then the queen is near where they are all pointing. With help of this kind it is much easier to find the queen than one might imagine.

Beekeepers cannot help loving their queen bees, and the queen must be an object of awe to anyone who understands what she is. She is at once beautiful and singularly stupid, a prisoner of her own fantastically specialized function. A swarm of bees is only imperfectly appreciated if the queen is not spotted within it, for she is its heart and soul. Twenty thousand swarming bees divested of their queen are without significance from the standpoint of nature, while a swarm with its queen seems to manifest nature's deepest workings and mysteries. This perception was expressed to me once by the landowner to whom I pay my rental fees with an occasional bottle of whiskey. He emerged from his house as I was hiving a swarm in his yard, having first gotten a few nips under his belt. We both watched spellbound as the throng of bees began pouring into the hive and I wordlessly pointed to the queen among them. He exclaimed: "And to think some folks say there ain't no Almighty." I do not suppose a swarm of bees rationally proves anything of theological significance, yet it did seem to me that my friend had found precisely the words to express the wonderment of the quiet spectacle before us.

EIGHT

Friends

A beekeeper has a chance to make lots of friends. I am not speaking now only of the human kind, who are not always the most constant, but of the countless others that relieve what could otherwise be an awful loneliness in the bee yards, woods and fields. We tend to take the song of birds, the squeak of the chipmunk and the scurrying of the red squirrel for granted. We may go hours without noticing them. But how desolate our world would be without them, and how desolate our lives! The hum of the bees lifts the spirits, but it is not enough. We need the whole of nature, and we need to be reminded that we are a part of it. The same life that pulsates in us, the same yearning and striving, the same love of existence fills everything around us. These things are not foreign, they belong to us and we to them. It is not our role as human beings to conquer, exploit and destroy, but to build up, protect and love in the spirit of acceptance the natural order into which we have been placed.

These thoughts are never far from me, and I was forcibly reminded of them one day when I began making preparations to move one of my yards to a more promising location. Beneath the old boards and odds and ends that I throw in front of the hives to keep the weeds down, and under many of the hives themselves, I

found numerous snakes, curled up in what they thought was the inviolable safety of those retreats. They were not attracted there by the bees, for I have never heard of any snakes having the slightest interest in a beehive. They were simply secure and, I suppose, warm places. It gave me a good feeling to discover them there, to learn that some of my hives and stands were serving this unintended purpose, and to know that, whatever may have been the snakes' terror at being so suddenly exposed to the light and impending destruction, they were perfectly safe from me. After a moment of that fixed stare and flagellating tongue, inimitable by any other animal, they slid casually into the surrounding weeds. I was sorry to cart off their roofs, but I needed them too.

Once, years ago, I was astonished to find a hen nesting beneath one of my hives. She had found a place safe from the entire world, protected as it was by legions of bees who would have no interest in harming her chicks. I managed to do my bee work there without disturbing her, and I have no doubt her chicks were hatched and brought into the world in good shape, although I was not at hand to witness this.

The kingbirds come around to my home yard, and these are officially described in beekeeping literature as enemies of bees. I suppose they are, in the strictest sense, for they eat them—they drop down and pluck them right from the air, near the hives. But are they not entitled to? As we rob the hives of their honey,

the kingbirds rob them of a bee now and then, and the right to do either seems hardly less violable than the other. I do not know how many bees one of the birds may eat. Perhaps quite a few if one were to number them singly. But certainly the number is insignificant in terms of the teeming population that a hive of bees supports. In any case, it is quite certain that the kingbird takes no more than he needs. Can we always say the same for ourselves? I have no objection to the kingbirds, and if my hives, in addition to the bounty they yield to their keeper, give up also a bit of nourishment to these graceful birds, then I have one more reason to be glad that I have them. Certainly, the kingbirds have never reduced my honey crop so much as a pound. They perch along the utility wire that runs across my front yard, spending much more time there than in the pursuit of bees, and in the few minutes they are perched there, far more bees are emerging from the brood combs in the hives than all these birds could catch in an entire afternoon. They follow their ancient patterns, as nature prompts them.

Of course the thought crosses my mind: would a kingbird hesitate to snatch a queen bee departing or returning on her nuptial flight, thus imperiling an entire colony? But then I realize that the likelihood of such an event is so infinitesimal as to be hardly worth the moment it takes to consider it. One could probably keep bees for a thousand years before it would happen.

Skunks visit the hives at night, scratch at the entrances, then snatch the bees that appear in the darkness in response to the disturbance. I find the image of this droll. I have never seen it happen, and probably never will, but more than once my dogs have gone boldly forth in the night to investigate and then returned to throw the household into shocked dismay, the skunk having known exactly how to deal with that threat. Still, I do not know what it would be like to live in the country without that singular odor of the skunk in the air from time to time, so I count its source among my friends. I can get on with him, and he with me; we simply keep our distances. Anyway, I have always thought that the Almighty was waxing whimsical when He created this remarkable animal and its extraordinary defense. The skunk can have a few bees and I shall not mind. I induced the bees to make their home in my hives, but I did not create them. As things are reckoned by men I own them, but in the order of nature I own them no more than the skunk does. The man whose thought would here turn to traps is still infinitely far from that sense of nature and spirit of acceptance I have spoken of.

Mice are a different story, and it would not be hard to learn to hate them if it were not such a perfectly simple matter to avoid their depredations. Various species of mice, a few of them quite beautiful, find an empty beehive a perfect haven. An occupied hive suits them in late fall, after the bees have clustered and become too torpid to put up a defense. The mouse chews out a hollow, ruining many valuable combs in the process, and stuffs it with the sort of trash it considers appropriate for a nest. A beekeeper should blame himself for this, however, and not the mouse, for it takes nothing more than a wedge of wire screening in the entrance to protect the hive completely. The mesh is large enough for the bees to pass through easily but too small for the mouse. Unless they become trapped inside and have no other way to escape, mice never, as far as I have been able to discover, chew holes in the hive. Nor do they ever climb up the side of the hive to take advantage of openings higher up. Of course, if, as in my yards, unused covers and bottom boards are kept in the apiary, to be at hand when needed, then mouse nests will frequently be found in them, often with baby mice. There is no harm in this. The worst that can happen is delay in the use of the equipment until the tenants have abandoned it. There is no need to dump the tiny creatures out to perish. Whatever task required the use of that equipment can usually wait until the next trip around.

Toads appear once in a while. It is impossible to think of them as enemies, although I have seen them included in that extensive list of "enemies of bees" in one of the leading treatises on apiculture. They may menace hives in tropical climates wherever hives are not raised from the ground, but certainly they pose no threat in more temperate areas. If they ever catch bees, which I doubt, they must get only a few of them. The toad squats Buddhalike, with boundless patience, certain that some sort of insect will come within reach of his tongue before starvation approaches. He impresses my own less trusting mind. I usually move him to my garden and hope he will stay there, but if he prefers to lumber back to the apiary I do not mind.

My friends are beyond numbering. At one yard, but only at one, the indigo buntings come around year after year. I have singled out these finches as special friends because of the indelible impression one of them made upon me when I was a boy. It was the first I had ever seen. Its beauty was exquisite, and I treated it as messenger of hope.

A woodchuck, too, comes around. He once squared off against my dog, who fortunately had the good sense not to draw too close. The 'chuck gets fatter and fatter until, by late summer, his sleek and portly bulk seems more to roll over the ground than to walk. I believe he is the champion hibernator, far exceeding the bear in his capacity for sleep. By the time he has sunk into his deepest torpor and begun the slow consumption of that great layer of fat his pulse will have slowed, I am told, to only four beats per minute. Then, according to quaint legend, as spring enters, he will rouse himself for the sole purpose of examining his shadow.

Even the crickets fit into the cycle, filling the meadow with their stridence at harvest time. It is hard to disassociate them in my imagination from heavy combs of bright new honey. They are drawn inside to the warmth of my potbelly stove when October's chill puts a frost on the grass, and there they add their singing to the otherwise somber autumn days. The stove seems to affect them in much the same way it does me. It holds the winter's sleep at bay just a bit longer and rekindles fading life. As winter begins to replace the fall I sometimes move a few of the crickets from my honey house stove to my cottage, there to make their home in a tiny bamboo box, sing by my bedside, and gain a further reprieve from winter.

I shall never understand nature, this earth, the bees, the buntings—all the myriad forms. No one ever will. I have no need to. I gaze in unuttered reverence, and I am fulfilled.

NINE

The Honey Flow

The time of honey flow is the one of quiet excitement for a beekeeper. It fills him with a sense of culmination, of a goal near fulfillment. Swarming is past, the supering is done, and the labor of the harvest will soon begin. In this interlude I can sit out among my bees in the home yard and hear the hum of their industry, smell the sweet redolence of incoming nectar and speculate upon its source, estimate my impending crop and gauge my riches.

Paradoxically, the period of the honey flow is the most relaxed time of the season. The activity is feverish, unremitting, but it is the bees that are working. Their owner is waiting, watching the weather, noting the blossoms, timing the opening of the bloom, keeping an eye on the scale hive and nourishing hope. All this is in vain, for there are no reliable signs. I can only wait and see how the flow develops. Still, no beekeeper can help watching for signs, however illusory. I note whether the basswoods have bloomed yet, and if they have, whether there are bees in the flowers. I check the patches of alfalfa, look to see whether bees are on the sumac bobs, note the flight of the bees at the hive entrances.

But the only real test is to peer into a few hives to see how the top supers are filling. By this time almost every hive has three supers, some four, and since the top one went on last it will be the last to be worked in. If there is honey in that one there usually is more in those underneath it.

I am always astonished by the volume of honey the bees store up in a short time during a good flow. A bee can carry only a drop of nectar and must usually fetch that from a considerable distance, sometimes from a source miles away. Furthermore, the nectar is more than half water and must be distilled before any honey results. Knowing these things, I repeatedly under-estimate the speed with which supers can sometimes fill up. I arrive at one of my yards to find the supers full and the bees needing more in a hurry. No matter how much nectar is in the fields, it will not end up in the hives or add a drop to my crop or a cent to my purse if the bees have no place to store it.

Having supers fill up with honey faster than empty ones can be prepared is sometimes a real problem, but it is also the nicest problem a beekeeper can have. When I push my hive tool under the crown board and find that it is well stuck down, I can sometimes sense that I am going to see lots of newly stored honey underneath it. The crown board comes off, and before my satisfied eyes are thousands of gentle bees, oblivious to this intrusion, and white combs bulging with honey. Bits of burr comb have been broken by my entry and the bees gather to lick up the pure honey, keeping everything dry and neat. Where all this honey comes from, how the bees could have gathered it so fast, each carrying so tiny a load, is a mystery. We sometimes imagine we know all the answers, but we never do, and even a beekeeper with a lifetime of experience is still often surprised. If the honey that drips from that broken burr comb has a deep mahogany color, however, and if it is late August, then I know where it came from, for this tells me that the buckwheat flow I so fervently hoped for has materialized and I am suddenly rich. I do not think I could find a surer happiness.

In the intense hum of the yards during a flow the bees are quite literally working themselves to death. Those that emerge just prior to the flow will perish after about six weeks, having done what their instinct has driven them to do to insure the colony's survival. Of course you do not get this impression of desperate, driving work when you visit the bees. All you see are

bees coming and going in great numbers, perhaps alighting a bit heavily at the hive entrance. Their morale is high and they do not sting me even though I may be sitting only a foot or two away. They are fulfilling their purpose without inhibition. There is no need to sting, no inclination to relax. If a bad turn in the weather at this point forces them to stop collecting nectar, allowing them to relax and extend their individual lives by an enforced rest, then they become irritable, which in their case means they become prone to sting.

I can tell when a honey flow begins by finding a dramatic rise in the readings of my scale hive. This is the special hive that rests permanently on a set of platform scales in my home yard. But this does not tell me where the honey is coming from, which is often a mystery. I am, as often as not, unable to find any conspicuous sources. But of course the sources can be inconspicuous. Unless examined closely, sumac blossoms do not even look like blossoms to us, but the bees assess them differently. One may observe dozens of bees moving from bob to bob of a sumac, touching for only the briefest instant each of the tiny flowers clustered together. The nectar gathered from such a blossom is microscopic, but to form the true picture of things one must multiply that molecular bit by the millions of such blossoms that are displayed and the thousands upon thousands of bees foraging over them for miles around. Other blossoms, such as basswoods, are similarly inconspicuous, except, perhaps, to beekeepers, who are always looking out for such things. It is not, after all, for our pleasure that flowers bloom; they bloom to attract insects.

Still other sources may be distant, their existence unknown to the beekeeper but perfectly known to the bees. It is universally claimed in the literature of apiculture that bees rarely fly more than two miles from their hive, and ordinarily much less, but I have reason to think that greatly extended flights are not uncommon. I have often found buckwheat honey in my hives, sometimes in great quantity, when I was quite certain there were no fields of buckwheat within two miles. Snow white fields of buckwheat, unlike basswoods or sumacs, are not nectar sources I could easily overlook, for I am constantly searching them out. Normally, there is not much else in bloom in the late summer when the buckwheat blossoms open, and the bees must either fly to those fields, at whatever distance, or not forage at all.

Although this is the period of the most intense and driving activity of the bees, it is, strangely and admirably, not competitive. The cooperation within the colony, which has been the foundation of the bees' survival over millions of years, appears to be carried over to the fields as well. Thirty hives sitting side by side exhibit not the slightest sign of rivalry for the bounty—or the scarcity—of the fields, each carrying on its work as though the others did not exist. Of course this mutual acceptance can suddenly terminate when special adverse conditions arise. If a colony becomes seriously weakened, for example, and the weather or advance of the season brings the honey flow to an end, a strong colony might "rob out" the weak one, taking every drop of its stored honey and leaving it to starve. This is rare, however, and is in fact almost always brought about by the bumbling of the beekeeper himself. If a worker bee enters the wrong hive, which is a fairly common occurrence when there are similar hives placed side by side, inhabitants of that hive are likely to attack it. Drones, on the other hand, can fly from hive to hive at will, and a worker that happens to arrive at the threshold of the wrong hive laden with honey or pollen will usually be admitted.

This spirit of cooperation, the avoidance of debilitating

rivalry, carried over even to the foraging areas, is one of the foundations of apiculture. Bees are never found fighting over discovered sources of nectar. Bees from a dozen hives can fill a single basswood tree without the slightest hint of conflict or rivalry. Two worker bees from different hives, which would fall into immediate combat if they were to encounter each other at the entrance to one of their hives, meet on a sumac blossom with only the barest sign of recognition. Under no circumstances do they race against each other to gather the nectar. Instead, one of them simply moves off to a different blossom, leaving the immediate field to the other. More than this, a given source of nectar never seems to be visited by more bees than can effectively and efficiently gather from it, even during periods when nectar is scarce. Thus the number of bees gathering from a patch of alfalfa or the blossoms of a fruit tree is proportionate to the abundance of nectar available there. As this diminishes so does the number of bees; they do not fall to fighting over the dwindling supply. If the source is a field of buckwheat it is visited by great numbers of bees in the morning, when the nectar is flowing, but by midday, the number diminishes to none.

When we marvel at the prodigious feats wrought by the bees, when we are staggered to incomprehension by the vast quantities of honey that can be stored by a single hive, we are likely to overlook what is surely part of the explanation: that the energy of the bees is directed entirely to this result rather than being spent in competition. Cooperation and mutual restraint are of course not unknown among other animals, nor among men, but I believe that nowhere else in creation does it approach this level, extending beyond the hive to the distant fields. To this the bees owe their survival; to this the beekeeper owes his harvest.

TEN

Getting Honey

To speak of *the* honey flow is misleading, suggesting that there is some identifiable time of the year when the bees gather honey, comparable to the flow of sap from maples. A honey flow results whenever any honey plant blooms in sufficient abundance and the weather conditions bring about nectar secretion. Both conditions are essential. Sometimes an entire field will bloom in beckoning profusion, yet be entirely empty of bees because the weather conditions are, in some undetectable way, imperfect. Thus there can be a succession of flows, waxing and waning, some heavy and others light, some long and some brief and some overlapping. They are seldom the same from one season to another, although in a given area there are likely to be one or more identifiable *primary* flows each year.

The earliest flows, from dandelions, fruit blooms and other spring sources, are used by the bees to build up their populations. These "spring buildup" flows in my area are sometimes followed by flows from the locusts, basswood, sumac, clover, alfalfa and other flowers. If things go very well, later on there may be a flow from buckwheat. Then finally in September come the goldenrods and asters, which under favorable conditions of moisture and warm weather produce heavy "fall flows," putting

the hives in condition for winter. The season comes to a close with the first frost early in October.

Sometimes a few days can be critical, and whether or not I get honey from a given source can be determined very quickly. The locusts do not bloom very long, for example, and many years they hardly bloom at all. If during the few days they are in bloom it rains on just the wrong one, confining the bees to their hives and washing the nectar from the blossoms, then there may be no locust honey at all. In the fall a shower a week early or a week late, in relation to the goldenrod, or a few days that are too cold or too hot, can make the difference between supers that are full and supers nearly empty. And so it is through the seasons. It is all in the hands of the gods from week to week. It is very rare for every major source to fail, but it is also rare for conditions to be favorable even for most of them. It is doubtful, for example, whether I shall ever in the same season garner good crops from locusts, basswoods and buckwheat, but I would be distressed if in any season I did not get one of these.

During a heavy flow, particularly at the beginning of the season when the work of the bees is most feverish, their drive is so great that nothing can divert them. I can leave my honey house door wide open, which would at other times be an invitation to disaster, and the bees ignore the odor of sticky cappings and combs. They are intent upon the more abundant offerings of the fields. Usually during these heavy flows the days are hot and humid and the nights warm, and every sunset finds the hives settled into a soft murmur, audible at some distance. This arises from the thousands upon thousands of wings fanning and miraculously evaporating and distilling the nectar into the golden richness of honey.

My sets of platform scales, each large and rugged enough to support a hive of bees the year round, year after year, are a source of joy, timing and measuring the flows, telling me at a glance what the bees are doing from day to day. I have two sets of these scales, one for a yard in the north, the other for my cottage yard to the south. Most of my other yards are scattered between these. The scales stay outside all winter each covered with only a bag, which is likely to be torn to pieces by squirrels looking for something for their winter beds. Still the scales continue to tell me what I want to know, season after season. Sometimes the gain is thirty or forty pounds in a few days, sometimes a dozen pounds from one sunrise to sunset.

When I multiply figures like that by my total number of hives and realize that I am the owner of approximately *that* much more honey, I think of the investor watching the rise and fall of the value of his holdings, his wealth varying from hour to hour; except I would rather watch the rising and falling of my platform scales, set out in the shade of a tree, than watch a tape or board in some windowless room. The scale hives provide one of those little extras. Each morning during the warm months, as I watch the various blooms unfold in succession and speculate over breakfast upon what the day will hold, I look forward to joggling that little brass weight down the scale to see where the balance will be struck that day.

Although two or more honey plants are commonly in bloom at the same time, the bees tend to gather from only one of them—whichever yields nectar more copiously. The pear trees may be in bloom and contain nectar in their blossoms, for example, but the bees of a given colony, and in fact all the colonies in the apiary, are likely to ignore them in favor of the dandelions that usually bloom at the same time. This creates a difficulty in the pollination of pears. The minor sources of nectar, such as certain garden flowers that do not bloom in profusion, are likely to be passed over by the bees completely if major sources are available to them. This disappoints beekeepers who have planned their flower gardens partly with the bees in mind.

On the other hand, this selectivity of bees is of great value, not only to farmers in terms of pollination, but also to beekeepers in enabling them to keep various honeys more or less separate. Ordinarily, if a honey flow has been strong, supers are likely to be filled exclusively, or nearly so, with honey from that source. If the honey is harvested at regular intervals as the season progresses there is a succession of honeys that are predominantly of a given type. There is likely to be a considerable mixture of the early honeys, but these are usually all light, mild ones that blend well anyway, so there is little point in trying to separate them. Every effort should be made, however, to get the honey harvested and empty supers on the hives before the buckwheat blooms, as this dark and distinctive honey is exceedingly valuable if kept separate from the others. The fall honey, mostly goldenrod, should also be kept separate, although it blends very nicely with the early light honeys. In these considerations we find the answer to a question that is frequently put to me by my customers: how I control what flowers the bees visit, thereby keeping the various honeys unmingled. Of course I have no such control, but the bees do, and I simply collaborate with their responses to the flows.

Honey supers can be either regular hive bodies, in which case they are called "full-depth supers," or they can be less deep, in which case they are called "shallows." A beekeeper with an unbreakable back who is assured he will suffer no loss of strength as his years increase can use the full-depth supers to advantage. He can interchange them with hives—with brood and food chambers—and he needs to handle only one super, whereas the beekeeper using shallows must handle two.

These advantages are, however, more than offset by the considerations that no man's back is unbreakable and even beekeepers grow older. When full, a mere shallow super is heavy, weighing forty pounds or more. Deep supers, when filled, are ponderous beyond practical limit. Of course, if one has a helper or hive-loading machinery, it is different, but this only calls attention to the disadvantage all over again—one is committed to hiring a helper or investing in such machinery.

Beyond this single, overwhelming consideration in favor of shallow supers, there are others that are not insignificant. It is, for example, much easier to keep honeys from different sources separate by using shallows. A deep super, taking twice as long to fill, is far more likely to contain a mixture from successive flows. The shallow frames, too, are easier to uncap and extract, and they are much less likely to contain honey that has not been capped over and is therefore "unripe." I decided upon shallows long ago and the wisdom of my choice has been confirmed many times. Beekeeping is hard, back-straining work, especially in the yards at harvest time. There is little to be gained in making it more laborious.

To facilitate uncapping them, only eight frames are used in the shallow supers instead of the normal ten. The importance of this can hardly be exaggerated. The combs are thick and can be uncapped in a single deliberate stroke. If foundation has been used, then nine frames are needed, for with fewer frames the bees build the combs irregularly and sometimes between frames. Once the foundation has been drawn and the honey harvested from them, however, eight is the proper number.

The principle behind these observations is that of the so-called bee space, a principle of utter simplicity discovered by the great American beekeeper Lorenzo Langstroth. It revolutionized beekeeping to the point of making genuinely commercial beekeeping possible. The principle is that bees can be absolutely depended upon to leave a space of one-fourth to three-eighths of an inch everywhere in the hive, and hence between combs. Given a larger space, they will build comb in it, while smaller spaces will be chinked with propolis. If eight drawn combs are more or less regularly spaced in a super, then the bees draw them out to the point where they are separated by the bee space and then cap them. This gives the beekeeper precisely the right comb to permit easy uncapping, as well as a good harvest of beeswax from the cappings.

Any beekeeper loves to have supers of beautiful white or near-white combs for the bees to fill; but things in nature are seldom that perfect, and with the passage of years the combs never stay white. The extracting combs become darkened, especially after brood has been reared in them, as it almost always is. Sometimes honey granulates in the combs before they can be spun out, in which case it is likely to be there the next year when it is time for the supers to go back to the hives. Eventually wax moths find some of the combs in storage, so that supers sometimes get wax worm webs between the combs. How does one deal with problems of this sort?

Dark combs are no problem. The color of the comb has nothing to do with the color, taste or quality of the honey that is stored in it. I have been using some of my extracting combs for twenty years. Over that time they have become dark, but they are still perfectly good combs, strong, well formed, and easy to uncap. The honey that comes from them is indistinguishable from the rest. When returned to the hives each season, they acquire the snow-white patina of wax at the commencement of the flow, and the cappings are, of course, renewed each year. There is no need to cull them merely because they are dark. The time do do that is when they have become broken, irregular or warped so they do not uncap easily.

Granulated honey in the combs, although a common problem, is also not a serious one. The bees eventually clean it out, but it may take more than one season. So here again, as in so many aspects of beekeeping, the bees solve the problems and do the work, restoring to use what would otherwise be of little value to the beekeeper, and correcting what are often the results of his own blunders.

Wax worms in the combs are similarly dealt with by the bees, so long as the combs themselves have not been broken down. The bees quickly remove every trace of the web before using the combs for honey storage. Of course there is a limit here to what one can expect of the bees. If the combs themselves are damaged and some of the moth larvae have spun cocoons here and there, then the frames must be cleaned out and fitted with fresh foundation.

Some apiarists have worked out ingenious ways to get extracted combs free from all stickiness before storing them for the winter. One can, for example, stack empty shallow supers, a half dozen at a time, over the crown board of a hive with only a tiny hole about the size of a pencil for the bees to pass through. The bees soon scour the combs of every trace of honey without storing fresh honey there or clustering in the supers, provided the hole for their passage is kept very small. There is no need for any more complex arrangement to achieve this result. I nevertheless always store my supers away sticky and return them that way to the hives the next year. I do this because it is easier, because I have seen other highly successful beekeepers do it and because winters where I live are sufficiently long and cold to prevent the residual honey from fermenting. There is another advantage. When the supers are returned to the hives in the spring the bees move right up into them without pausing a moment. They love those sticky combs, take them right over, clean them up at once and go right to work in them.

But I must confess there is also a disadvantage to this. The sticky combs in my open truck are frightfully attractive to the bees in my home yard, especially after the first spring flows. So I have to get everything in readiness first, back the truck up to my honey house, load up the supers quickly and be off and on my way before the bees realize what is going on. Otherwise, the whole place becomes bedlam in a matter of minutes, and my dear wife again presents me with the now-familiar ultimatum: either her and my happy home or this mad obsession with bees. It would be terrible to be forced into actually having to make a choice, for it would be impossible to abandon either.

ELEVEN

Puttering

Beekeeping charms the naturalist, the philosopher and the putterer. If one happens to be all three the blessings are uncountable. The bees and their astonishing order open the eyes of the naturalist to what others behold without seeing. The philosopher is able to view mankind in the context of a far greater creation. He sees nature not merely as a stage for the unfolding of human history but as a cosmos, a great thing of beauty (which is what the word meant to the Greeks) in which every facet of creation has its place and purpose.

We are not, however, just beings who observe and think. Our joys are sometimes found, not in the contemplation of wonders, but in our fingertips, in doing things, inventing, fabricating or, to use the best word of all, in puttering. Here the child in us, which the poet declared is the father of us all, finds his playful outlets. Here is what is sometimes needed to make us come alive—to be concocting schemes, seeking out little things, preparing surprises for ourselves, testing our own cleverness, sometimes experiencing success and then stepping back to look at some trivial thing we have worked up and thinking: Lo, what I have wrought!

I do not think I have ever known a serious beekeeper who was not a putterer. It is one of the great joys of the craft, and when beekeepers get together for their annual regional conference there is almost certain to be, alongside the honey and beeswax displays, a special gadget show, with premiums awarded to the cleverest and best. The creators of these are usually near at hand to explain them to any who show the slightest sign of interest. This attention to gadgetry is a little surprising when one considers that the basic equipment and paraphernalia of apiculture have not changed much in a century. I sometimes find in old catalogs of bee supplies, put out at about the turn of the century, many of the same things that are shown in catalogs today. It would appear that the tools of the craft have become perfected and fixed, but not so. Every beekeeper introduces his own variations and improvements and, inevitably, his own inventions, things hitherto unknown that he—and perhaps he alone—finds indispensable.

Under the head of puttering I shall include all the activity ancillary to beekeeping that has to do with tools and equipment, whether they are things invented by the beekeeper or familiar things picked up and sometimes put to novel use. This latter is extremely common. Indeed, the really passionate beekeeper can never pass a junk shop without going in to see what he might find to use in his tool box or honey house —something to use in melting beeswax, an old burner for making steam, a useful scraper, whatever. Thus did I acquire my lovely old potbelly stove for a few dollars, attracted as much by its beauty as by its anticipated utility. By now it has become the soul of my honey house, transforming it into a place of comfort and warmth. My honey house is filled with such things, ranging from a fine brass fire extinguisher I found in the road and recharged, good as new, to all sorts of pans, and dippers, strainers, mixers and spoons. It would be almost impossible to inventory them all. Many were made up out of my head, some picked up here and there. I make it a rule, however, not to collect junk, and as soon as I discover that I am not using some bit of equipment, out it goes.

Still, there is plenty left, and the overhead beams of my honey house are covered with nails from which hang these dozens of bewildering objects. Things that cannot be suspended from nails have their places on the many shelves erected here and there. These are themselves old bookshelves and cabinets that someone had thrown out. It is seldom that some tool I need is not in sight and close at hand. My eye is constantly open for things I can bring home. Even when the reward is nothing more than a pile of baler twine or some old burlap bags that a farmer has cast aside, still, they will fuel my smoker for many weeks to come.

For several summers farm auctions were a rich source of gadgets. When commercial farming began replacing family farms not long ago there were left behind many farmhouses and buildings, still inhabited, but not by families engaged in farming. Over the years many of these were abandoned and their contents auctioned off in the yard. Attending these auctions, I always hoped that some past tenant had been a beekeeper, and that his things, as well as old books on beekeeping, would now be unearthed. And sometimes the hope was rewarded. It was in this manner that I acquired some of my best smokers, sometimes offered by auctioneers under such strange names as "bee lamps," and as unfamiliar to the bidders as to him. One auctioneer was glad to let me have a considerable stack of hive bodies for five cents each, there being no other beekeepers in the

audience to offer ten. It was through farm auctions that I got two splendid sets of platform scales, one for a dollar, the other for seven. Both have been of inestimable value in helping me plan my work and schedule my harvests.

These auctions have other rewards, too. At one I found a fine old bee book, of no particular monetary value, and one I already had. Having nothing else to wait around for, I moved it to where the auctioneer would pick it up, thinking I might offer fifty cents. This gesture of mine was noted by an antique dealer, however, who interpreted it as a sly one and inferred that I had discovered something of rare value. Soon she and another dealer in collectors' items were bidding against each other furiously until one of them triumphed with an outrageous bid. I believe I enjoyed this display of uninformed greed more than I would have enjoyed owning the book, which I would only have given to a friend.

My most memorable auction came as a total surprise. I drove off to it with no special expectation and saw nothing about it to stir any interest, until suddenly my eye fell on a bundle of top bars to frames, neatly tied, dusty, but unused. That was all —nothing else to suggest bee things. It was like finding a diamond in a coal yard. How did these get there? Could there be more of this stuff somewhere about? There had to be, it would seem. But aside from this one tantalizing item I could find no evidence of any bee equipment anywhere on the place. I went poking all over, but in vain. Deeply puzzled, I returned to where I had discovered the top bars and probed deeper. Suddenly my heart stopped as there was revealed, beneath a miscellaneous pile, a mountain of bee supplies—box after box of valuable foundation, frames, a copper steam generator of just the kind I needed, and evidently lots more, all old, dusty, but mostly unused, judging from what I could see.

Now I began to eye the crowd to see whether any of the others looked like beekeepers. One gentleman who seemed interested in my discovery pointed out the little steam generator and informed me that it was an "extractor." I decided there would be little competition there. The auctioneer drew closer, and I assumed the appropriate uninterested expression, concealing the excitement and anticipation within. Just as it was the business of the auctioneer to get bids up as high as he could, it was mine to hold them down as best I could, and soon I was loading up my truck on a winning bid of eleven dollars. It was a memorable moment. Not until I got home and took inventory did I discover the true extent of my treasures. It was a good boost to my honey business and gave me many good hours of assembling and wiring frames in the winter months that followed.

Nothing, however, really equals the things you make up for yourself, particularly when they are the product of your own creative imagination. Most such inventions are utterly simple, requiring neither inventive power in their conception nor skill in their creation. A scrap of plywood, for example, is used to press lids onto pails, saving my hands, and it hangs on a nail to be used each year. A five-gallon can with a tiny can soldered on for a spout is used to refine beeswax to perfection. Labels need not be pasted one at a time and applied to containers. Instead, I paint paste onto a wide pine board, spread a couple dozen labels on this, then lift them from the board to containers in a rapid and orderly way. That is my labelling machine, which cost only the price of the paint brush—its only moving part. It never gets out of whack, unlike the labeling machine offered in the bee supply catalog for several hundred dollars.

Stacks of supers weighing several hundred pounds are rolled around in my honey house on little dollies. The dolly consists of a plywood piece with a rim tacked to the top and a ball bearing. tacked under each corner. At first I screwed casters to the corners. This required drilling sixteen holes and turning sixteen screws. That was all right. But roller bearings, I have discovered, can be purchased for eighty-nine cents for a set of four and can be fixed to the dolly with only a few blows of a hammer, as each comes mounted in a casing with a nail. This simple improvement greatly enlarged my honey house, for with more than a dozen of these dollies on hand I can move things about and make space with little effort.

My hive stands, mentioned earlier, are another lesson in simplicity. Each is made by cutting a ten-foot two-by-four into two lengths of three and a half and two of one and a half and nailing these into a rectangle to hold two hives. These cost virtually nothing and completely meet the need.

Some beekeepers with a special skill for woodworking make all their hives and parts, including frames. Quite apart from my lack of skill to do this, it does not seem worthwhile. Frames are of precise construction, and the factories are able to turn them out accurately by the thousands. I have also found that hive bodies can usually be purchased second hand for less than the wood would cost to make them. Covers and bottom boards, on the other hand, are another matter, for they can be made out of scraps. A cover is made by nailing one or two pieces of plywood to a rectangular frame and covering it with an aluminum sheet. Such sheets can be had for almost nothing from most newspaper plants, where they are used once and discarded, a fact that seems to be unknown to most beekeepers. Bottom boards are just as simple. Factory-made ones are of complex construction and, for some unfathomable reason, reversible; that is, they provide a shallower space on one side than on the other. This reversibility is worthless, for no beekeeper is going to remove a whole hive in order to flip the bottom over. My own bottom boards are easily made by nailing rails to wide pine boards laid crosswise, then painting them on the outside with creosote.

It is always a challenge to invent ways of speeding operations in the honey house. The shed attached to mine for super storage also serves as a honey-warming room. Full supers from the yards are rolled in on their dollies on cool fall days and left there for a day or two while a small electric floor heater, controlled by a thermostat, brings them to the temperature of a warm summer day. This greatly speeds the job of spinning them out. Sometimes your inventions can save you not only time but much overhead expense as well. An example of this is the frame rack I just recently invented. I was about to buy a second extractor for several hundred dollars, so that my uncapping procedure could proceed without interruption, one extractor spinning the honey out while I filled the other. But just then I came across an enormous saucer-shaped reflector, about three feet across, which, with a bit of figuring and scheming, I was able to suspend from the ceiling by a pipe rigged with a roller bearing. I load this, turning it slowly as I go, while the extractor spins, thus solving my problem at almost no cost.

Invention and the exercise of creative genius mostly fills the winter months. The summer's work exposes the bottlenecks, the procedures where motions are wasted, and the winter's woolgathering supplies the means of overcoming them. On cold nights, warmed by my stove or bed, or sometimes when trips or other monotonous pursuits force me into the company of nothing but my own thoughts, I can concoct new schemes or devices. There is never a dearth of material for my thoughts, and I approach each season with the conviction that, however splendidly things worked for me last year, they are going to go even better this time.

TWELVE

Comb Honey

Raising comb honey is one of the most ancient of agricultural crafts. In fact, until fairly recently in human history all honey was either comb honey or honey that had been crudely crushed from the combs. Even what would now be called the modern methods of production have hardly changed in nearly a century.

Comb honey is what honey truly is. The honey we see in shops, in neat uniform jars, is not exactly what the bees made. Rather it is what people have taken from what the bees made, the "improvement" human beings have imposed. Here, however, it is almost incontestable that nothing has been improved, except in terms of convenience, that good comb honey represents a kind of perfection never found outside of nature and rarely equalled by nature itself. Nothing tastes better, and what is perhaps as important, nothing is more beautiful to the eye. Who has fully sampled life's joys who has not tasted clear, white comb honey on a biscuit? What beekeeper has really tasted gladness who has not taken comb honey from his own hives? And what is nearer pure fulfillment than garnering one's crop and seeing the sections of snow-white combs grow into a veritable mountain? It seems almost a pity to see them sold. No rich man gazing at his gold could be happier, for comb honey is more beautiful to look at and is the fruit, not of greed, but of hours of beekeeping under the clear sky.

I shall never forget the first time I saw comb honey, and how it fascinated me as a small boy. My mother brought it to our cottage and let me cut it from its little wooden box, explaining that the bees had made it for me. I have few memories that extend back so far. The construction—and of course the loveliness of the transparent honey that ran from the broken cells —must have charmed me. I do not remember. Perhaps it was at that moment I became a beekeeper in spirit. Raising comb honey does, in any case, seem to me the culmination of the art of apiculture.

Comb honey is simply the filled and finished honeycomb that the bees have constructed in little boxes supplied by their owner. These boxes of honey have come to be called "sections," from the manner in which they fit into supers. Some older people still refer to them as "cards," for some reason unknown to me. Comb honey is not essentially different from the fresh honeycomb one might take from a bee tree. It is, accordingly, the only genuinely natural honey there is and, by that token, the only natural sweet known to mankind. All other honey has had something done to it between harvest and use, has had its nature altered, however slightly. The change is usually not great, but it is perceptible to a beekeeper. Even honey that has been only spun from the combs and strained has lost something that is discernible to the taste. Therefore, comb honey is to me the specimen of perfection, the ideal from which one may depart for practical considerations, but which cannot be duplicated.

Despite all this, comb honey is not the first honey choice of most people. Partly this is due to widespread ignorance of what it is, even among people who love honey, and partly to its higher cost. No doubt some people also feel that eating comb honey involves eating beeswax, which they find difficult to think of as food. Actually, because the bees draw it out to such unbelievable tenuousness, there is little wax in high-quality comb honey. Those who are accustomed to it usually swear to its wholesome effects, and I certainly add my voice to theirs.

Virtually all the comb honey I raise, which is a considerable amount, is sold directly at my roadside stand. I have sometimes sold it to stores and roadside fruit stands, but never with the great success I have had at my own stand, where nothing is sold except honey. Many who stop pick it up out of sheer curiosity: they have never seen it before. Some express skepticism at my label, which declares that the bees themselves put it in the containers. Many others are elderly people who were familiar with comb honey long ago but seldom see it now. I sometimes feel that I am fighting a lonely battle for wider use of this loveliest of foods. Children, I have noticed, invariably love it, and having discovered it they will probably seek it out for the rest of their lives.

For the beekeeper the great advantage of raising comb honey, and one that seems insufficiently appreciated, is that no special equipment is needed. The only tools are the regular apiary tools—smoker, hive scraper and so on—plus a pocket knife. One does not even need a honey house. No one in his right mind would spin out honey in the kitchen, but there is no problem in putting up comb honey there. A prodigious amount of equipment is needed for strained honey—uncapping equipment, extractor, tanks, strainers, steam generators, the honey house itself, and so on. In addition, spinning out the honey and packing it is considerable work.

Nearly all of this is avoided by the comb honey apiarist. He simply takes the loaded comb supers to the kitchen or shop, empties them, scrapes unwanted beeswax or propolis from the sections, and they are ready. The two great advantages of producing strained honey are that the honeycombs can be returned to the bees year after year instead of being consumed as part of the crop, and honey that has granulated can be restored to a liquid by warming it. Nothing can be done to restore granulated comb honey, and while such granulation does not spoil its value as food, it does detract from its desirability. A beekeeper must, therefore, sell his comb honey crop promptly.

I believe the best of everything is achieved by raising both kinds of honey, unless one has only a few hives, in which case there is no need to raise anything but comb honey. My comb honey crop is the first to come off the hives, so I can begin selling it at my roadside stand at once, along with jars of honey saved from the previous season. Later, when the buckwheat fields bloom, I try for a second large harvest, blending the rare goodness of buckwheat to the natural goodness of comb honey.

The early honey, from the locust trees, clovers and basswoods, is light and beautiful in the white translucent comb. It is gathered rapidly at this time of the year, when both days and nights are warm, so the resulting sections are well capped over. I can also get this honey off my hives before the heavy work of harvesting the extracting honey commences, thereby adding pleasure to the summer days. Many aspects of beekeeping hold special joys, but this one is especially great—to sit out in my neat little honey house, with my ancient radio keeping me company, and in a leisurely fashion put up hundreds of sections of comb honey, using no tool other than my pocket knife. The honey then moves right from my honey house to my deep freezer (a step to be explained in due course) and then to my honey stand, where it is eagerly snapped up by passing motorists.

Beekeepers tend to think that comb honey getting requires extraordinary art and patience, perhaps more than they possess. Non-beekeepers, on the other hand, imagine that it must be relatively easy, that it differs from getting strained honey only in that the last step, removing the honey from the combs, is omitted. Actually, there are certain procedures and manipulations special to raising comb honey, but there is no special art beyond the basic understanding of bees that any beekeeper should possess.

Getting comb honey requires somewhat more attention to the hives. When extracting supers have been put on hives they can be disregarded for a while, sometimes for weeks, and it will not matter—they will fill up with honey sooner or later and the bee master can then garner it at his convenience. One must, however, keep a close eye on the comb supers, for they must be gotten off the hives almost as soon as they are filled. If they are not, then the honeycomb becomes darkened, or "travel stained," as it is called; that is, the cappings lose their whiteness as a result of the thousands of tiny bee feet continuously walking upon them. This does not hurt the honey in the least, but it detracts from its beauty.

It is important that comb honey supers be filled rapidly, during the early or midsummer honey flows when the weather is warm. The fall flows should not be used for comb honey, other than to complete the unfinished sections left over from the earlier flows, for comb honey made during cool weather is waxy, less delicate and, if it contains goldenrod, granulates quickly.

There are many ways to raise comb honey, some simple, some not so simple, and some, of course, better than others. I think I have tried them all, and I shall describe two that certainly work. One is exceedingly simple and requires virtually no skill; the other is more complex but, when properly done, produces stunning results.

The simple way is just to set comb supers over one's strongest colonies, with no more special attention than that. While this involves the least work, it is not the best system. Those strong colonies would produce more extracting honey than comb. They are also more likely to swarm with comb supers on them. Still, one who is raising comb honey for the first time cannot do better than to use this method, and it is still one I use each year on at least some of my hives. Sometimes, for example, when I run out of extracting supers and a good flow is on, I pile on comb supers rather than have hives in the yard with no supers at all.

The other method involves some radical manipulations. I have never seen it fail, however, provided the special manipulations are correctly done, so it is no doubt worth the extra time and work. The system is as follows. Early in the season, when the dandelions are blooming and before the main honey flows begin, remove a strong colony from its bottom board and set it down on a new bottom board immediately in back of where it was, facing in the opposite direction. On its original bottom board, still at the original stand, set an ordinary shallow extracting super fitted with nine frames of foundation—not drawn comb, but foundation *only*. Now find the queen in the original colony and set the frame she is on, queen and bees and all, to one side for a few minutes out of the way and in the shade. She will be safe there and will not fly away.

Next, *shake* most of the bees from most of the remaining frames of the original colony in front of the shallow super on the original hive stand. Do this until about two-thirds of the hive population has been shaken. This procedure has already been described as "shook" swarming, and it is a well-known method of swarm control. Before this "shook" operation is completed, remove the queen from her frame, snip off about half of one wing (or the tips of both wings if it is easier) and let her run into the shallow super, along with the other bees that have been shaken off there. She could, of course, simply be shaken off with the rest, but then one would not have a chance to clip her wing, and it would also be easy to lose track of her.

The bulk of the population, together with its queen, has now been transferred to the new shallow hive. Add a queen excluder, plus three comb supers, cover both hives in the usual way, with crown boards and hive covers, and give the bees the rest of the day to settle down. It is desirable, although not essential, to introduce a new laying queen to the original, depleted colony on the new stand, as this will enable the colony to rebuild its population with great speed and get in condition to gather a fine crop of honey itself.

Consider, now, the result of the foregoing manipulations.

The original hive, although otherwise intact, has lost its queen and most of its population and finds itself at a new location. Either it has a new queen, supplied by the beekeeper, or it immediately prepares to raise one. For a few days there will be little activity at the entrance of this hive, for nearly all the flying bees will either have been shaken at the old location, where the shallow hive now sits, or will have returned to it of their own choice, since this is what they consider to be their hive location. But in a couple of weeks this original hive will again be populous, for brood will have emerged in large numbers and bees will again be foraging. Meanwhile, there is no chance whatsoever that this colony will swarm, because it not only has a new queen, but more significantly, it has lost its flying bees. By the time this colony regains its strength it loses all interest in swarming.

On the original stand there now sits a shallow extracting super with nine frames of foundation, and over this the excluder and three comb supers. The shallow super becomes the new but temporary brood chamber for this colony. It is as if the bees had swarmed and taken over this shallow hive. They occupy the supers at once, since these, like the shallow hive below, contain foundation. This is the secret of the system's success. The bees immediately begin drawing the foundation and storing honey in the supers. If the beekeeper peers into the supers after a week has passed he will find comb honey being made in all three at once, and at an astonishing speed.

This new colony must be watched closely for the first day, however, for it is likely to swarm out either the day the shook operation was performed or the next morning. There, is, of course, no brood at all in the hive to hold the bees there, and it is as though they resented the whole procedure and were determined to begin afresh in a place of their own choosing. They will not go far, however, because their queen is clipped. If the bees swarm, the beekeeper need only find the queen in the grass in front of the hive, place her at its entrance and let her run back in. This done, the bees quickly return on their own. Then they settle down to establishing a new colony and, of course, making comb honey. There is not much room for them to store honey below, for there is only a shallow super that rapidly fills with brood as the queen resumes egg laying. Virtually all of it, therefore, goes up above, soon to be taken by the beekeeper.

When the heaviest early honey flows are over and most or all of the comb honey has been garnered, the two hives are reunited in the following manner. The hive that was moved, and which has by now grown to enormous strength under the new queen, is put back on its original stand, and the shallow hive, now without any comb supers but with a huge population of bees, is placed on top of it, where it becomes an extracting super like any other. All the brood in it hatches out and it becomes filled with honey. The flying bees from the moved colony soon find their way around to the new entrance and are accepted there. The colony has two queens for a while, of course, but one of them, most likely the younger one, soon deposes the other. This colony, uniting the populations of both, now produces a good crop of extracting honey from the fall flows, and of course it does not swarm, for the swarming season is long past.

The system is well worth the extra labor it involves, for, if done properly, it not only results in a great honey yield but also requeens the colony and puts it in excellent condition for winter. The one disadvantage is that it leaves the beekeeper at the end of the season with new extracting supers filled with darkened comb. This is a disadvantage unless one needs additional extracting supers. Thus, if you manipulate ten colonies each year in the manner described you add ten dark extracting supers to your collection each year, whether you need them or not.

There is however, a way of avoiding this, and that is to use a full-depth super instead of a shallow for the shook swarm. This of course results in additional full-depth supers every year, but you are likely to need these anyway for normal replacement and expansion, and it does not matter in the least that the combs are darkened. The result of using full-depth supers instead of shallows is not so spectacular in terms of production, but you are likely to have fewer headaches in the long run.

Those are the two best methods of getting comb honey. They are not described in the manuals of beekeeping. In fact, most manuals merely instruct you to reduce the colony to a single story, crowding the bees and thus "forcing" them up into the comb supers. Not one of them, so far as I know, describes just how this is to be done. Most beekeepers know, too, that bees are not easily forced to do anything, and that merely crowding them in a hive, with no other precautions, is the best way in the world to precipitate swarming.

I have tried the following system, however: divide a strong colony, leaving the queen with the *heaviest* of the two stories on the original hive stand, and add comb honey supers, one at a time, to that. Here the problem arises of what to do with the story that is moved. I have simply piled them up together, four high, on new stands, let them build up to strong skyscraper colonies, added supers as needed, and then in the fall distributed them back to the hives they had come from. But this was not a good system. The bees were slow in getting to work in the comb supers, the drawn comb in the single brood chamber attracted them more. A number of hives also swarmed, spoiling everything. And the four-story colonies made up of the stories left over were very inefficient honey producers. The bees had so much room to story honey that they were not interested in putting it into supers. On the basis of all this I do not recommend the method of hive reduction.

There has been one highly significant innovation in comb honey equipment in the last several years, and that is the substitution of round plastic sections for the traditional square sections made by folding strips of precut basswood. These round sections are superior in every way, but the equipment has, unfortunately, not been properly marketed and is no longer easy to get. For the past twenty years all of my comb honey has been produced in round sections.

A form of comb honey that has become increasingly popular in the past several years is the so-called *cut* comb honey. This is raised in regular shallow extracting frames, using special un-wired comb honey foundation, then cut from the frames in uniform rectangular shapes. This comb honey is, of course, as good as any other and much cheaper to produce, although all the frames must be cleaned each year to be used again.

About the only special problem in producing comb honey, of whatever kind, is the damage caused by the so-called lesser wax moth. The tiny larvae of this small moth puncture the cappings of comb honey, riddling the surface of the comb with tiny holes in rather short time. Evidently, the parent moths lay their eggs in cracks around the outside of the hive and the minute worms then emerge and work their way into the supers. So long as the super is on the hive the little wax worms can do no significant damage, but the moment it is removed and the honey is no longer under the protection of the vigilant bees the worms can roam about undisturbed. They do no real damage to the honey, but they ruin its appearance, and this is serious enough.

Beekeepers once dealt with this problem by fumigating the comb honey with bisulfide while it was still in the supers. It would be hard to imagine a worse fumigant. Bisulfide is foul smelling, highly flammable, and toxic to the beekeeper himself if inhaled. Yet it did not affect wax worm eggs, so the supers had to be fumigated a second time to destroy newly hatched larvae.

Fortunately, this disagreeable procedure is no longer called for. It has been discovered that the wax worms are destroyed in every stage—egg, larval and adult—by extreme cold. Thus I pack my comb sections promptly after removing them from the supers, then wait about three days for any microscopic eggs to hatch. This wait is not essential, but it does guarantee the results of the cold treatment that follows: piling the sections into a deep freezer, about a hundred at a time, and letting the temperature drop to five or ten degrees below zero (F), which takes a day or so. This produces no change whatever in the honey, but it does destroy any vestige of wax moth life. The freezer is then turned off and the honey left inside for two or three days, by which time it has returned almost to room temperature. It it were removed while still cold, moisture would condense on the sections, perhaps to harmful effect. Using this simple operation, which is still not widely known, I can raise large crops of comb honey without any damage from wax worms.

Someday I shall be too old to maintain far-flung bee yards and to haul the tons of honey to my honey house, and too old to spin it all out, for this is heavy work. But it is hard to imagine being too old to raise comb honey. The work here is sometimes more time consuming, in relation to the amount of honey produced, but it is seldom arduous. So I can speak of someday retiring from commercial beekeeping, but it is not easy to entertain the idea of retiring altogether as long as the art of comb honey getting remains and as long as I can use a simple pocket knife. Considering that this art has been cultivated by men and women, and very often by elderly men and women, almost from the beginnings of agriculture, it is not likely to pass away.

THIRTEEN

My Honey House

The beekeeper's precious retreat is his honey house, but unlike a bee yard, this curious structure and its contents are a constant invitation to intrusion. People assume that its owner could not possibly be doing anything of importance in there and feel quite free to come and go as they please, some to ask questions, some to buy honey, but most just to talk. Sometimes, of course, they are welcome, but most of the time I want to be alone there. When I am in my honey house it is likely to be either the scene of intense activity, which I can ill afford to interrupt, or a place for puttering and pondering, which one can hardly do in concert with others. Beekeeping stimulates more profound woolgathering and concocting of schemes than any pursuit I know of. To do it properly and fruitfully, one must be solitary, developing a detached and serene spirit.

My own beloved honey house, unlike so many of the outbuildings that evolve to that use at the hands of beekeepers, was actually built to be a honey house, according to my own specifications and plans. It cost rather little, since it is no more than a neat frame structure erected on a concrete base, but over the years much has been added inside and out. It now wants very little either in beauty or utility. Flower boxes grace the windows, and each spring I put petunias in them. At the same windows green shutters, picked up at the village dump and repaired and repainted, give the structure a homey look. Morning glories climb to the roof each summer on one side, and just beyond these my herb and vegetable garden blooms and fruits. On the other side the birds nest in the now overgrown honeysuckles, and an aged clematis spreads itself over the side of the adjoining shed that I built onto the honey house for super storage. Of course the structure is too small—eighteen by twelve feet, with the attached shed for super storage measuring only six by twelve. But everything is compact and well organized, so that I am able to do what I want to do quite efficiently.

One end is my shop, where I repair hives, clean things up, make things, and in a word, putter. A beekeeper needs rather few woodworking tools, which for me is fortunate, as I have a limited skill with anything more complex than a hammer or saw. Still, necessity does enlarge one's creative powers, and I have sometimes stood in astonishment to see what my hands have wrought. There is, for example, a fine little buzz saw, made up from an arbor I picked up for two dollars, a blade, a discarded kitchen table and an old washing machine motor. It is adequate for whatever extensive sawing is required. In my shop there are cigar boxes by the dozen, odds and ends of other containers I have picked up, whose contents and accumulations of the years are known to me alone and would be unfathomable to the rest of the world. Old book shelves and cabinets provide storage for assorted tools and implements, soldering materials, and a fairly large collection of old smokers, most of which need repair.

It is a nice place to be. My friendly radio keeps me company

when I feel the need of it, and can of course be silenced the moment that need ceases. My little potbelly stove is there, acquired as a luxury, I thought at the time, but it quickly proved its worth. It burns wood and banishes the chill of a fall day in minutes, not only warming the shop and its proprietor, but lifting my spirits as well.

The other end of my honey house, which takes up a bit more than half of the total space, is devoted entirely to equipment for spinning out, straining, storing and bottling honey, and to molding the beeswax resulting from this operation. As tanks fill with honey, small plastic pails fill from the drizzle of melted wax, resulting in neat stacks of wax blocks as the honey spinning progresses over the course of many days. The wax is worth far more than the honey, pound for pound, but the amount of it, although considerable, is small in comparison to the tons of honey.

The honey house goes through a regular cycle as the year advances, its operations harmonizing with the orderly stages of work in the yards, which is rather the same every season. The cycle begins in the spring, when the hundreds of supers stacked away in the shed are distributed over the course of several weeks to the hives. Each hive eventually receives three supers —perhaps only two for a few backward ones, but sometimes four for what I call the "busters." These do not go on all at once. The second super goes on when the bees have begun to fill the first in earnest, then the third when the second has begun to fill. Everything must be done in an orderly way, taking account of the development of the colonies and the waxing and waning of the early honey flows.

Not much planning is needed here, however. My truck and trailer are loaded up with supers and these are distributed to the yards each weekend, as weather permits, and at the same time I deal with spring cleaning and swarming. It all works out about right. The combs are restored to the care of the bees before the wax moths return. By the time they get filled with honey, spun out and returned to the shed the season for the destructive wax moths will have passed.

After supering comes time for tidying up, restoring everything to order after it has been used and more or less dismantled the previous fall. With all my equipment back in shape, clean and shiny as a new penny, tanks and other containers covered with clean white cloths, I can step back and contemplate my orderly little plant, all ready to go on a new crop. Of course not for another year are things going to look quite that neat and orderly again.

The interlude between supering up and harvest is the time for repairing equipment and getting together those things you need. Each summer sees innovations and improvements, largely the product of my woolgathering. Frames broken and other pieces of equipment are taken care of, and my inventive schemes are carried out. It is a pleaseant interlude, a relaxed one in which I can set my own schedule, and a valuable one.

I have always taken a deep satisfaction in putting to use something that has been broken or discarded as useless. Old frames salvaged from a hive whose combs have been riddled by wax worms or mice can be cleaned up with lye and put back to use as good as ever, thereby enlarging my honey crop without unnecessary expenditure. Old hives and supers, many showing the marks of considerable age, can be cleaned up and restored. Odds and ends can be fashioned to whatever purpose my imagination has conceived—a new idea for a strainer, perhaps, or a rack for extracting frames. The possibilities seem endless as the years come and go. Now too is when the comb honey is harvested and made ready, when accumulations of beeswax scraps are melted down, and when the crystallized honey saved from the previous season is reliquefied and bottled to get my roadside stand into business ahead of the harvest. There is no rush here, no pressure, and whenever my chief impulse is simply to sit out with my bees, watching them come and go, I can indulge that too.

The operations in my shop and honey house are governed by three rules. The first is: No bees. They are rigorously excluded as not belonging there. Bees that get in are promptly removed.

A few bees in the honey house are quite harmless, having no inclination to sting, but they nevertheless are a distraction.

The second rule is: No honey drippings on the floor. This rule requires special vigilance when combs are being spun out, but it is important, for general neatness as well as for ease of work. It is a constant distraction to step in honey, then track it about, stepping in it again. A couple of good big sponges and buckets of water solve this problem.

And the third rule is: No people. I think I love people almost as much as I love bees, but I do not love either species in my honey house. Under no circumstances may any caller be told, as was once common, "You'll find him down in the honey house." Instead, the visitor is announced to me there, and I decide whether to come or to have the caller sent on his way. It depends on how engrossed I am at the moment and also, of course, on who the caller is.

There had always been a tacit understanding to this effect. My dear wife, for example, had long ago learned to knock tentatively when she came to the honey house door, knowing that an intrusion there does not always produce a cheerful greeting. But the rule was only loosely enforced until one disastrous July afternoon.

I was at work spinning out honey. All was going along well until my friend Osmo, a recent convert to apiculture, appeared at the door bearing the single super of honey from his single hive. In an expansive moment I had offered to spin this out for him. My honey house set-up would inspire questions from anyone of normal curiosity, but in trying to attend to the flow of Osmo's questions, I became completely distracted. I apparently stepped on the gas hose that feeds the burner to my steam boiler, shutting it off, unnoticed, and thereby setting off an unbelievable chain of events, unknown to me until the discovery of their appalling consequences later on.

No sooner had Osmo left and my work resumed, than I was dumbfounded to behold in the doorway the smiling face of the president of the university where I occasionally teach. He had chosen just this moment to see whether I might like to go sailing. I nearly dropped a heavy super on my foot. My venerable president, coming upon me as extractor, honey pump and uncapping machine were all going at once, with a great whirring and clatter, remarked that I had a regular factory there. (I could not help wondering whether perhaps he was thinking that I could be using my time far better than at such folly as this.) (The next day, however, I got to thinking that harvesting a nice crop of honey is not half as frivolous as sailing around in boats.)

By the time I had shut down the motors to make conversation, chatted amiably, declined the sailing invitation and seen my second visitor on his way, the smell of gas led me to have a look under the steam boiler. Well! Because the burner had been inadvertently shut off when Osmo called, the boiler, in cooling, had sucked melted beeswax out of the pot into which the vented steam had been exhausting, drawn it all the way up the hose, through the uncapping machine, on through the rest of the hose and into the boiler itself! I could not believe my eyes! It took me a good hour to clear the system, and by this time I decided to leave the job for another day. Meanwhile, my third rule was delivered unto my family, with Mosaic finality and to the accompaniment of thunderbolts overhead: No people in my honey house—ever!

These are precious recollections. They bring a smile, even though the episode seemed horrendous at the time. Life is not always simple and easy—it would be monotonous if it were

—and perhaps the good memories one can store away, of varied kinds, are really more lasting and more genuine than bank notes. Certainly, the more they are used the brighter they get, and they do not ever really get spent.

FOURTEEN

Stings

A beekeeper worth the name develops a healthy contempt for stings. Indeed, he really should start out with that attitude before he has received his first one, so that this aspect of his craft can always be kept in proper perspective. Over the course of years I have woven a veritable philosophy around my attitude of sublime indifference to stings. I have needed the fortification of this philosophy, for a sting does still hurt, just as much, I think, as when I was a novice beekeeper. And it still brings from my lips the same sputtered oaths. In spite of this, I have to maintain a genuine and stoical lack of concern for them. I have to really believe, when I am working with the bees, that I am *not* going to be stung. This is the central requirement for the demeanor necessary in the presence of bees. I have to believe this even though I know, if I pause to think about it, that it is the exception rather than the rule if I conclude my work in a yard with no stings.

Bee work requires intense concentration. One has to keep one's mind on the task at hand, without distraction. Otherwise, blunders are made and the bees might, in fact, become antagonized, so that instead of facing the possibility of a sting or two one is suddenly met with the threat of hundreds of them. You can avoid such a sudden turn of events by keeping your mind on what you are doing. Success in dealing with bees demands a certain demeanor. What is called for is efficient and deliberate movement. It is not so much *slow* motion that is wanted but a controlled approach. Such control is a difficult thing to describe, or even to demonstrate, but nothing destroys it more quickly than anxiety and fretfulness. The moment you jump in alarm you abandon the temperament essential to beekeeping. Some people possess that temperament by birth, as a part of their nature. They somehow sense exactly what is called for and the bees do, in fact, tend to keep their peace with them. Others, although they may have kept bees for years, never develop a deliberate approach, and their hours in the bee yards are hectic and disorganized. Stings are their appropriate reward. These persons should never have taken up the craft, and few of this kind ever do. They belong in the wrestling ring rather than in the apiary.

The best description of the demeanor needed for beekeeping was conveyed to me years ago quite by accident. A sweet and saintly woman came upon me as I was working with some hives, and she was impressed by my bare hands and shirtsleeves. After watching from a safe distance for a while she remarked: "You just send love out to them, don't you?" That is it exactly. It is not just a matter of loving bees; I suppose every beekeeper loves bees in some sense or other. It is more a thing of spirit or attitude. However absurd it may sound to those of scientific orientation, a good beekeeper sends love out to the bees, without giving it any particular thought. In that frame of mind, the work goes well, smoothly, efficiently, without upsets and, in fact, usually without many stings.

Despite what most people believe, bees are not prone to sting. Cross bees are the exception, gentle bees the rule. Most wild things are gentle, and bees are too. They are among the gentlest things on earth, far superior to men in this respect. They have their own purpose, as all things do, and it is not to sting. They must be provoked into it. A beekeeper who stands in the midst of his apiary, perhaps clad in nothing but shorts and sandals as thousands of bees fill the air around him, is not

being courageous. His mere presence is no signal to the bees to attack. For years I kept a dozen hives on my garage roof in the heart of a city, with neighbors and children on every side, and there was never a single complaint of a sting. I was finally obliged to move them, not because anyone was stung, but because a new neighbor moved in down the street, and, noticing beehives, became alarmed at the thought of stings and complained. At that point, upon moving the bees, I moved myself as well; for although I loved my house, I loved my bees more.

The bee was given her sting for the defense of the colony and nothing else. Hence, bees attack almost entirely in proximity to the hive; elsewhere they can be goaded to sting only by extreme measures, such as being stepped on with a bare foot.

The same is true of hornets and wasps. The sting is a defense mechanism. Nature is indifferent to the loss of this or that individual insect and makes no provision for its survival, but the destruction of the colony threatens the species itself. It is there that the life of the species is carried forward. Of course, virtually every sting received by a member of the general populace, whether in proximity to a beehive or not, is called a *bee* sting, so bees, innocent of any involvement in 90 percent of such incidents, nevertheless suffer condemnation for them all. There is hardly a large lawn or meadow that does not harbor yellow jackets, which go peacefully along unnoticed, their nest concealed in the ground, until someone disturbs it. Then the report is spread of a vicious stinging by "bees," and the incident is likely to be so reported in the local newspaper if it is serious. Proper identification is rarely made. I have even seen large paper hornet nests referred to as "beehives," and more than once I have been summoned, in my capacity as beekeeper, to come deal with them.

It is in this manner, too, that bees have acquired the reputation of being more dangerous than rattle snakes, a comparison I have seen solemnly reported in newspapers. More persons, it has been discovered, die from insect stings than from snakebites. This is, in fact, true, because a minuscule part of the general population is hypersensitive to insect stings. For these few, a sting is a real threat to life. Despite precautions, these people are likely to receive a sting sooner or later. The likelihood of a given individual ever suffering a snakebite, on the other hand, is relatively small. Rather few people ever see a poisonous snake, much less receive a bite from one. These basic facts are then modified into the falsehood that more people die each year from bee stings than from snakebites. This, in turn, becomes the outrageous general statement that bees are more dangerous than rattle snakes, which makes splendid newspaper copy.

The misconception might be harmless enough if it did not also separate vast numbers of people from an appreciation of the most beautiful aspects of nature, not to mention the strange reputation it imparts to beekeepers. It does sometimes make life difficult. I have already mentioned my having to move the apiary from my garage roof, not as a result of mischief done, but simply mischief ignorantly feared. The same kind of fear nearly cost me my best apiary site. I found what seemed to me a very promising place to set up a yard and approached the farmhouse a quarter of a mile away to seek permission. The farmer assented at once, but his wife was alarmed at the thought of bees visiting her garden. My reassurance that bees, after all, fly for miles only alarmed her more. She evidently thought they would fly such distances only to seek her out and attack her in her garden. It is pathetic that such wedges of fear and ignorance should be driven between people and nature, for we shall always be part of that nature, whether we understand her or not.

A beekeeper soon learns that bees are subject to swings of mood and temperament just as people are, and generally from the same basic causes. This is no projection of human feelings into other animals. It is one of the most obvious facts of apiculture. The general formula is very simple. When all is going well and the bees are able to satisfy their ends, they are agreeable, but when their efforts are frustrated, they become cross. A beekeeper can work with his bees in his shirtsleeves on a warm spring day when the life of the colony is expanding and acres upon acres of flowers are quite suddenly opening, offering pollen and nectar in greater abundance than the bees can take advantage of. Hives can be taken apart, combs removed for inspection and the bottom boards cleaned with only the slightest notice from the happy and industrious bees. Sometimes on such a day even the veil can be discarded with little risk.

When, on the other hand, the sources of nectar have dried up, as is likely in August, and the bees sense a threat to their well-being, they respond angrily to the slightest disturbance to their hives. Again, when the beekeeper has harvested honey, leaving little on the hives, the bees are sometimes reduced to a very ugly mood until more nectar is gathered and the threat of starvation is banished. Their mood swings from one extreme to another even in a single day. Many beekeepers have wondered, for example, why the hives are so cross when buckwheat is in bloom. In fact, they are not then cross in the morning; only in the afternoon. The explanation lies in a peculiarity of buckwheat itself. It secretes nectar only in the morning.

A beekeeper, knowing the signs of these swings of temperament, is often credited with great courage when no credit is due. Swarming bees, for example, are notoriously gentle, at least for the first day or so. They have abandoned their hive, therefore have no hive to protect and no inducement to sting. So some beekeepers do not even bother with a veil when they deal with a swarm. This always makes a stunning impression upon onlookers, and I have awed many audiences this way.

And of course I always have beehives near my house. I would not want to live where I could not. It is one of my pleasures to present visitors with a comb just taken from a hive and covered with hundreds of bees, among which they are invited to pick out the queen. The bees have no reason to fly from the comb and sting, and virtually all of them are young nurse bees anyway, but this performance is invariably met with stiffened expressions on the faces of my guests. If people will not open their eyes to the wonders and loveliness of nature, if they insist upon closing them in fear and distrust, then I feel that I must, when I can, force their eyes open. I feel like Plato's philosopher, entering the dark caves of ignorance in which people immure themselves and dragging them into the sunlight.

It was discovered in antiquity that smoke instantly mollifies the mounting anger of bees, transforming them from a bold and threatening army into a passive and retreating throng. This is a very useful fact of their psychology. A beekeeper would usually be helpless without his bee smoker. A gentle puff or two gives him complete control of the situation under almost any circumstances. No one seems to know why this is so, and I am often asked what the smoke "does" to the bees. I am convinced it does nothing to them, in any ordinary sense. Certainly, it does not harm them in any way. My view is that they react to smoke in exactly the way we do—they turn away from it. Their defense is, in any case, an appropriate one, for they do what would be the best thing to do if threatened by a forest fire. They turn their attention to their stored honey and make ready to salvage that, at least, in case all else should go up in flames.

It is a widespread belief among beekeepers, even among those who have studied bees for years, that upon sensing smoke the bees all gorge themselves with honey. Sometimes this is even offered as the explanation for their gentleness in the presence of smoke—that they are so filled with honey they can no longer bend their bodies in the manner necessary for implanting a sting! Which only shows that there is probably nothing too absurd to be believed by someone, even sometimes by experts. When a puff of smoke wafts over an open hive one can, indeed, see bees soon drawing honey from open cells, but only a minute fraction of the total population is so engaged. The rest go about their activity much as before. Certainly the hive as a whole does not "gorge" itself, and authorities who have suggested this have substituted imagination for vision. A bee filled with honey, is, moreover, still perfectly capable of stinging. She has simply lost the inclination.

There are varieties of stingless bees, but they are not of much interest because they gather no significant quantities of honey. And I am, on the whole, glad that our honeybees do sting. I find something to admire in the way they can bear down on an intruder, even when I am the intruder. Beehives in gardens would probably be as common as tomato vines if their occupants did not sting, and while this would be nice, in a way , it would also rob the beekeeper of that specialness that I, at least, prize. And it would, it seems to me, be an imbalance in nature if anything as delicious as honey were obtainable with little effort. As things are, the bees hold their own against the virtually omnipotent human species, and I am glad it is so.

FIFTEEN

The Cross Hive

Everyone can recall certain moments that were turning points in life. The moment of my first bee sting was surely one of these—not just because I had been stung, even though the immediate consequences of this turned out to be rather horrendous, but because the fates had somehow put me in the proper frame of mind for it. From that moment my attitude toward bees, and their stings, has been essentially what is required for the art of beekeeping.

I was still young but had already sensed that the hostility people see in nature is born of ignorance. My love for nature had turned to awe and amazement when I had recently beheld my first beehives. Soon afterward I went out to the country to call on an ancient beekeeper I had heard of. About the only thing I knew about bees at that point was that they will not sting if you do not bother them; and this was less an item of knowledge than an article of faith derived from my general attitude toward living things.

I found the beekeeper, very old and crippled by arthritis, and he took me at once to his apiary. There must have been a dozen hives. I had never seen so many bees. Striding dauntlessly up to them, in the steps of the old man, and like him without any veil, I did not wince as I felt a sting on the very top of my head. I somehow had the faith that the bees would not attack me if I approached them in the right spirit and that this sting was some sort of accident. I was right about that, for there were no more, but within a half hour my entire body, to my toes, was covered with brilliant red blisters, and I hardly made it home and into a bathtub before my eyes refused to open. For several days I wore a face that rivaled that of the ape. The experience seems only to have reinforced my determination to become a beekeeper, and in any case, those miserable initial effects have never been repeated. A sting now, even to the eyelid or lip, produces the same sharp pain and the same expletives, but no swelling.

Every old-time beekeeper can tell about the cross colony he once had that no power of heaven or earth could subdue. Such a colony is not usual, but is encountered sometimes. It violates every principle of apiculture. A puff of smoke only rouses it to greater fury. It is invariably a colony under stress—one that has just lost its queen, for example, or one that has been robbed of most of its honey.

My own encounter with such a hive is unforgettable, although I have to admit now that the anger of these bees was provoked by my choosing an inappropriate time, during a dearth of nectar in the fields, to take their honey. This hive did in any case teach me a lesson about that, and it dispelled conceit that with skill and the right approach I could manage any colony of bees on earth.

I arrived at daybreak in the yard where this colony stood, thinking I could get my work done before the heat of the day. The deep red of the rising sun warned that the day would be hot and therefore ill-chosen for work in a bee yard. And indeed it was hot, even before the sun was very high. When I saw a grasshopper in front of my nose, *inside* my veil, I should have interpreted it as an omen and left the work for another day. Things went all right for a while. I was using a bee blower and had about forty supers off the hives and loaded when off came the cover of the cross hive, which had never been particularly cross before. It was as if they had been observing my approach for the past hour with mounting fury.

Every beekeeper can imagine the scene that followed. The bees rose from the open hive in a cloud. My shirt, already plastered to my back with perspiration, was quickly stitched down as if by a thousand needles. A few bees even managed, miraculously, to pass through the wire mesh of my veil. This unsettles the nerves of the most stoical beekeeper. I got the hive back together, gathered up my equipment, and with elephantine grace got myself away from there, observed with profound curiosity by a young field hand not far away. Glancing over my shoulder as I fled, I noticed that the field was suddenly empty, the field hand having evidently moved with approximately the speed of light from the center of the field to its periphery. And sure enough, there he was, off to the edge, his shirt off now and twirling slowly, like a windmill, in front of him. I didn't ask any questions. My faithful old dog, who always accompanies me on my bee trips and who has never failed to respond to my call, did not appear this time. I became convinced that her brave heart had finally given out, that a thousand bees were now stinging her lifeless body in some forsaken place where she had vainly sought safety. Actually, she was eventually found, stretched out comfortably by a cool spring in a darkened cow barn nearby, oblivious to the fierce bedlam of the world outside.

That evening my dear wife was shocked to count the stings I had received. I could have told her where she might have found a

few more. The prettiest twist to the whole episode came about a
week later, however, when I returned to find this same colony
now fairly good-natured, and to learn that the young field hand,
whose eyes, I was told, had been swollen shut for two days, had
decided he wanted to raise bees himself, was eager to learn more
about them, and wanted me to give him lessons! Well, he
seemed to have gotten through the hardest lesson with good
grades. He moved away soon after, however, and our paths have
not crossed again. Perhaps he is now keeping bees.

I wrote to the bee supply company that night for a complete bee suit with zip-on veil, an item that had always been considered the necessary uniform by female apiarists of my acquaintance. This brought to a full circle my own philosophy about stings. That philosophy had begun with a scorn for even a veil and with the conviction that stings are the reward of the clumsy beekeeper. Perhaps one gets less heroic with age, but I generally use my bee suit now, with its zipper veil that no bee can penetrate and the ventilated leather gloves I bought at the same time. Now I do sometimes spend an entire afternoon with the bees without a single sting. Cowardly or not, it seems good.

Still, I know that as long as the bees are such a part of my life, which will surely be to the end of my days, there will be plenty more stings. It is worth only so much to try to avoid them.

SIXTEEN

Garnering Honey

In most branches of agriculture harvesting is the most reward-ing phase. It is the time to taste the fruit of one's labor and measures one's success. It is not always so in beekeeping, for the gathering is not a straightforward matter of reaping and taking. The bees, never tamed, do not always yield up what they have gathered with the docility of domesticated flocks. Yet an ex-perienced apiarist has learned their ways and knows how to steal away with their precious honey before they are even aware that he has done it. If their resentment is roused by this, then their owner, if he knows what he is about, is miles away by the time it happens.

The commonest mistake of inept beekeepers is to try taking the honey on a hot day when the flowers have dried up and the bees, already in a sour mood, are ready to take out their frustration on whatever moving thing draws near their hive. Sometimes, too, a careless beekeeper permits sticky honey to become exposed in the yard under these already stressful condi-tions, by breaking bits of honeycomb or by using a mechanical bee blower which requires breaking the supers apart before blowing the bees out, creating drizzles of honey here and there. This can create a frenzy of excitement in the bees. The air becomes filled with them, and no veil or bee suit is total protection against them. What should be a joyous task becomes irksome, nerve-racking toil, tempting novice beekeepers to wish they had never gotten into such business.

I harvest my comb honey supers as they become filled throughout the season, beginning in July. Escape screens, which permit the bees to move downward a few at a time into the hive below, but do not allow them to return, are inserted under the supers. Cleverly arranged little wires spring nearly shut behind them as they pass through, permitting only a one-way passage. After a day or two the supers are empty or nearly empty of bees, and I simply walk away with my bounty. Any bees still remaining are sent flying with a few good huffs and puffs as I walk off, for a comb honey super is not so heavy that I cannot hold it in front of my face and blow into it. At the same time I can savor the sight of the lovely comb sections and make a preliminary assessment of their quality. I do not think of comb honey simply as good food; it is the distilled nectar of flowers sealed into snow-white waxen cells of the most exquisite beauty.

The means of harvest is about the same for the honey that will be spun from the combs in my extractor. It does not matter how long it has been on the hives, for even though the cappings of the combs may have become darkened by the thousands of bees moving over them day after day, this makes no difference whatever in the honey that will be spun from them. When there has been a heavy flow from the locusts or other early sources I try to make a preliminary harvest early in the season, mainly to have it on hand for the hundreds of people who are eager to buy it. But there is no need to do this. If the honey flows are slow and drawn out, so that it takes most of the summer for the supers to fill properly, then the harvest can be put off to September and the honey will be just as good. In that case, however, it is harder to keep the different honeys separate.

By early September, the time of the last and largest harvest, the days are usually bright, pleasant and cool. The hives are tall and loaded with honey. Those that have three shallow supers on them will give me nearly a hundred pounds. Some will give me more, some less, but when I average a hundred pounds overall for the season I am content. This adds up to a ton of honey from an average yard of twenty colonies.

A day for harvesting is at hand, then. There will be more harvest days in other yards, and another day to return and finish the work at the yard I have chosen for this day. Every such day will be pleasant if early snow does not overtake me. I have learned the pitfalls to avoid, learned how to avoid stings, how to avoid straining my back and limbs, and how to avoid excessive perspiration. There will be time, as I proceed with the work, to appreciate the crisp air and bright sun that only autumn provides, and to relish the sight of the goldenrod and wild asters on every side. Sometimes where the soil is very rich the goldenrod grows higher than my head and I must pick my way through it to find the hives it conceals.

The bees have an air of contentment. Their hives are filled and they feel secure for the winter—a state of being that has been the purpose of all their gathering. They would feel less secure if they knew what is about to happen, but it will take another million years of evolutionary change to teach them the meaning of the beekeeper who now approaches. They long since learned the significance of the bear, whose name in Slavic languages means "honeyeater," but the methods of a beekeeper are more deft and subtle than the rampaging bear's. Their brains are no match for mine.

Of course, I do not menace the bees. I am going to take most of their honey, but they will still have more than enough to carry them through until spring when their colony life will begin anew. And it is not unfair that I take it either, for it is because of my skill and the means I have provided that they have been able to gather so much more than they need in the first place. Left to themselves in the hollow of some tree they would gather much less. I have earned what I shall take.

I go about the work with my veil tied snugly, my bee suit zipped up securely and my trousers tucked into my boots. My bee gloves, with their long gauntlets, are there if I need them, but I probably will not. It is doubtful whether my bare hands will receive more than one or two stings, possibly none at all, and the greater dexterity of bare hands is worth this insignificant risk. My smoker is fueled, smoldering well, and my pockets bulge with spare twine, corn cobs and burlap scraps for occasional refueling. Another pocket holds a roll of masking tape and scraps of newspaper, which I shall need before long.

Now to begin. I approach the first hive from behind and pry up the top two supers, leaving them tightly stuck together by the propolis the bees have applied to every crack. Just a few cool and gentle puffs of smoke tell the bees they should stay where they are, and they do. They are in no way harmed, but an instinct tells them to guard their precious stores and be prepared to fly off with them in case fire, suggested to them by the smoke, really threatens. Raising the two supers and drawing them toward me an inch or so, I tilt them forward with one hand and hold them effortlessly at a near balance for a moment while my other hand slips the escape screen under them. The two supers are then lowered onto the screen and the whole business is shoved squarely into place. A generous piece of asphalt shingle, softened slightly by the warm sun, is laid over the crown board hole where it fits snugly, keeping any bees from entering by that route. Now the job is virtually done for that hive, but the hive cover is left off for the time being while exactly the same procedure is repeated for every other hive in the yard, perhaps twenty or so in all. By that time I have slipped escape screens under a thousand or more pounds or honey, and yet I have done no strenuous lifting at all, only a brief balancing act at each hive.

Next I return to the hive I began with and scrutinize the supers above the escape screen for any crack or opening through which bees could enter and rob out my crop, which they will almost certainly do if they can. That is why the hive covers were left off: I want to be able to see that all crown boards, as well as the supers themselves, are perfectly bee-tight. There are always a few that need a bit of chinking and taping over: That is why I have the newspaper scraps and masking tape in my pocket. Where there are holes and cracks large enough for a single bee to pass through there are by this time several bees clustered, bees that have emerged from the supers. They are a dead giveaway, drawing my attention instantly to the leaks I am looking for. If I had new equipment these precautions would not be needed, but most of my supers are old, some are very old and weathered and cracked and chipped here and there. When I am satisfied that all is tight and bee-proof I replace the covers and the brick that belongs on each and the work is done for the day.

There is time to count the supers and calculate roughly the wealth they will add to my pocket. Forty supers will yield well over a half-ton of honey and many hundreds of dollars. A day or two later, or a week later if I am busy with other things, I return to gather up the supers, heavy with honey but empty of bees. I load them one at a time onto my truck and my little trailer and

cover them well with cloths and crown boards, in order to keep temptation away from the bees. The loading up can sometimes be done without even wearing a veil, for the theft is so swiftly performed that the bees do not even know what is going on.

The last super having been loaded, it is now time to gather the escape screens still on the hives, and here smoker and veil will be needed again. A billowy puff of smoke goes right down through the escape screen before I even pry it loose. The screen is removed and replaced by the crown board and cover. Bees still clustered on the escape screens are dislodged with a single sharp blow in front of the hives, and I am then ready to drive off with my crop.

Thus is the harvest accomplished, a different yard each week, sometimes two at a time, always with at least two such procedures at each yard. Meanwhile, I spin out the combs in my honey house and store them away, emptied of honey, ready to be used again next year.

It is good work, and the tons and tons of honey that are spun out through the warm fall days between trips to the yards restore both my body and my mind. The sight and fragrance of a hundred or more gallons all ready to be drawn off into glass jars is by itself enough to ease whatever effort it took to get them.

SEVENTEEN

Honey Spinning

Autumn has come and I am deep into spinning out honey. These October days are incomparable to any others of the year. They seem laden with the sense of ending, and yet mingled in this is a heavenly beauty. They carry the sad feeling of bidding farewell to beloved friends, and yet one sees that this is but a phase in the eternal cycle. The period of fruiting is over, woods and fields are transformed, and what remains of my garden's harvest sprawls over the ground in the morning dampness. A few slugs take advantage of the sun's slow retreat. The occasional geese overhead, sometimes close, sometimes distantly honking, add to the sense of departure. But who can be blind to the sweetness of a quiet fall day?

The flowers have dried up and offer nothing to the occasional visiting bee except, here and there, some late pollen. Nature's activity is coming to a halt, well ahead of the first frost, but the hives are heavy with the late fall honey. They will winter well this time, no doubt of it, and I need not worry on cold winter nights that my bees will be hungry, as I sometimes do when a cruel early frost can rob some of my hives of their chief hope of survival. The bees now come and go, as always, but they are listless, their flight mostly pointless, the product of sheer habit, as the months until spring stretch out before them.

Autumn always creeps up on me and then pounces, threatening to trap me with an early blanket of snow before my honey is harvested. It has never caught me, but it sometimes comes close, and each year I am driven by anxiety that the first snowfall will find me with extracting supers still on the hives. I am not sure that this would be a disaster or why it always seems so threatening. The honey could still be garnered even with snow on the ground. But everything would seem out of harmony, the mood would be wrong, and an activity that is normally filled with satisfaction would instead have the quality of an anxious dream. I always press, then, to have everything done before the first of November—the honey all spun out, the beeswax stacked in uniform yellow blocks, and the honey house itself tidied up and the stickiness rinsed away. Then with the first snow and howling wind I can feel that another phase of my beekeeping year is completed, while a new one begins, and I feel in step with the cycles of nature.

A day for spinning out the honey begins as I pull myself from my warm bed in the chill early hours. Soon the sun begins to dye the eastern horizon red. Before long, it will drive the heavy fog from the lake and then dry the wet grass. A bit of tea helps stir me to life as I think of the stacks and stacks of supers filled with honey and waiting for me in the honey house. They have been kept warm through the night in my little heated super shed, so the honey will flow as nicely as on a hot summer day. As dawn brightens I contemplate, over breakfast, the thin column of smoke spiraling from my honey house chimney, rising from my potbelly stove that I kindled last night and ignited in the sleepiness of the early morning. The honey house will be toasty

warm and smell of the apple wood cut from falling lir.
nearby ancient trees. Every year at this time a few crickets i.
their way in. I do not know how. The structure seems secur.
against ants. But the crickets respond to the potbelly stove in
about the same way I do, and they are certainly welcome to share
it with me. Their bright notes are just the spice that is needed
on these sometimes somber days.

The honey spinning continues until after supper and into the
night, but my fatigue is compensated by tanks of precious
honey. Besides this, a bed feels better to tired limbs than to a
body that has done nothing, and not many thoughts can equal
those that culminate a day of achievement. It is a lot of work,
sometimes monotonous work, and I am sure there are people in
other pursuits who, without working this hard, end their day
far richer. But I would not relinquish my small honey house
with its devices and old motors and gadgets, its simple machinery, its stove and its crickets. Once one has tasted the sublimity
of nature, heard the hum of the bees, beheld the migrating
geese, absorbed the song of the cricket and felt the intimations
of the glory of God, one has no great temptation to try burnish-
ing one's own greatness. The world and the hours, these are
enough.

My honey house setup borders on the fantastic, at least to the
eye of the layman, although it is all perfectly intelligible to me,
since I designed it. Every element in it is intended to save time
and work. It has evolved gradually over the years as new ideas
and inventions have been incorporated until now it seems nearly
perfect. Perhaps it resembles nothing so much as a one-man
band. The master of the apparatus stands at the center, operat-
ing various switches and motors with both hands and feet, and a
great and orderly process revolves mysteriously around him.

Reconstructing the setup now in my imagination I find two gas burners for the generation of steam, which I must keep an eye on for proper adjustment, three motors salvaged from old refrigerators and laundry appliances, a great quantity of switches to activate motors and warmers, floats, a warning

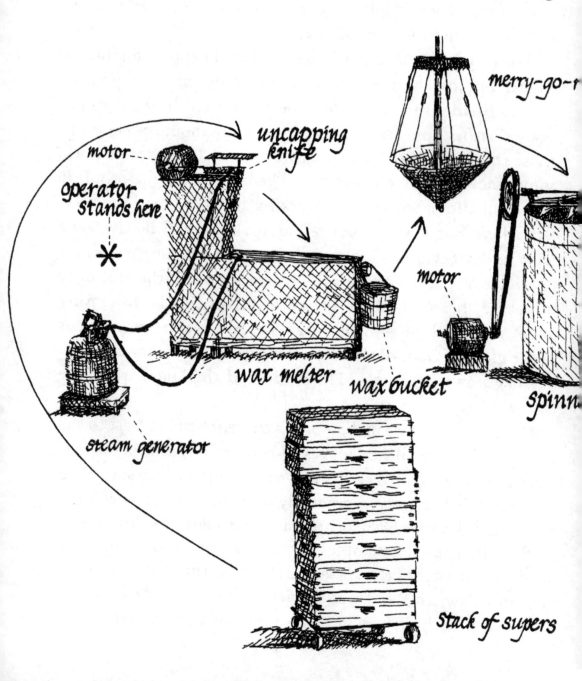

merry-go-r

uncapping knife

motor

operator stands here

motor

wax melter

wax bucket

spinn

steam generator

stack of supers

buzzer, various levers, paddles, and some miscellaneous things for which it would be hard to find a name—these in addition, of course, to the usual spinner (or extractor), tanks, uncapper and so on. The instruments of my one-man band look something like this:

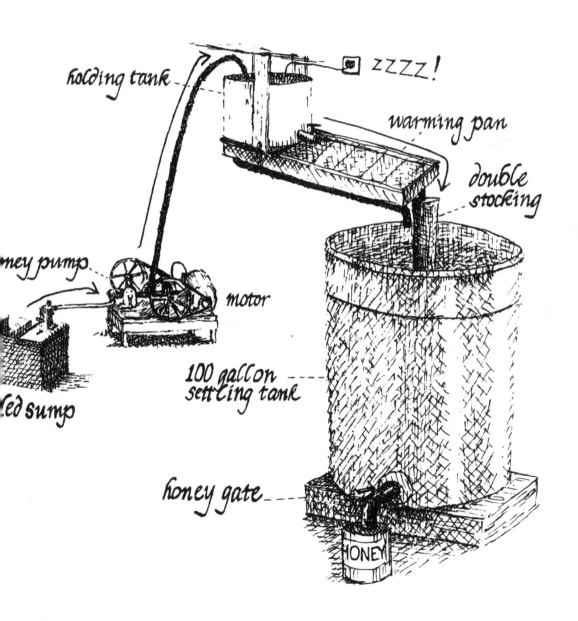

Of course items are represented in this picture as strung out in a line. It would be quite out of the question to try representing them as they really are, for the whole complex apparatus is arranged more or less in a circle. I operate from the center of this circle where the many things needing my attention are within reach. Here is how it all works.

The supers are stacked, eight to a pile, on dollies that are rolled about where needed, as described earlier. I never need to lift a full super once it is in the honey house. The combs are loosened by prying up one end of the super and slipping a special stick underneath, thereby raising all the frames for easy grasping when the super is lowered onto the stick. A steam-heated knife is kept in rapid vibration by a motor that is turned on and off by a special switch at my toe. This slices the cappings off with a single motion for each side of the comb. The uncapped combs go one by one into the spinner and, when this is filled and running, they go into what I call the merry-go-round. This is a revolving rack of my design that holds the combs until the spinner is again empty and ready to receive them. The cappings, meanwhile, fall stickily onto a hopper and flow from there into my melter. This is an ingenious device that imparts no heat to the honey that accumulates in it but melts the cappings that float on this honey by means of steam-filled copper grids on top. As the wax melts it drizzles off into plastic buckets that serve as molds. Other than to change the plastic bucket from time to time the melter requires no special attention once it is steamed up.

My honey spinner is made to hold twenty-four combs, but an additional twelve go in perfectly well. Even though some of the combs rest flat against each other the honey all spins out as it should—a fact that, I have found, is little known among beekeepers. They can greatly increase the capacity of their radial extractors by taking advantage of it. The slots of the spinner are marked with a pattern of colored dots, which enables me to distribute the combs of each super throughout the radius of the reel and then restore them, after spinning, to the supers from which they came. This not only insures that the combs do not get mixed up but, more important, that the reel is always balanced and will not wobble. There is a system to this. Each super is emptied beginning at the center, and the combs dropped into the spinner according to the colored dots—red for the first super, yellow for the next one, and so on. Thus, even though the combs are of different size and hence of different weight, and even though some may be partly granulated, the spinner remains balanced.

Ideally, two people should work together at spinning. The nature of the work is perfectly suited to such cooperation: one can uncap while the other loads, runs, and unloads the spinner. I nevertheless prefer to do the whole thing by myself. I can start and stop everything according to my own mood and convenience. While the spinner is whirling away, driven by its old faithful motor and belt, I turn to whatever needs attention —adjusting the flame under the steam generators, checking the beeswax drizzle, scraping burr comb where it has accumulated or, of course, uncapping more combs and loading the merry-go-round.

The honey pours forth with a rush the moment the spinner attains speed, then continues to flow at a decreasing rate as the combs rapidly empty. It is conveyed by gravity to a baffled sump, which retains most of the particles of beeswax that have been spun out. As the sump fills, it automatically raises a float activating the switch to my honey pump that is driven by the motor from a washing machine. The pump conveys the honey to an overhead tank, made by cutting an oil drum in two. This is the holding tank from which the honey flows at a constant rate into the warming pan beneath it. If the honey in this holding tank rises dangerously near to the top, so that it might overflow, a buzzer sounds, activated by a float in the tank, and then I need only to turn off the pump for a few minutes.

The warming pan that receives the honey from the holding tank is a long water-jacketed pan interlaced with baffle plates. An immersion heater and thermostat keep the water in it at a constant temperature. A temperature of 130° F. is usually sufficient to retard granulation for several weeks, but the device can be set above or below that as needed. The honey flows back and forth across this pan, guided by the baffles, then empties into a double-thickness nylon strainer consisting of two ladies' stockings, and then into the one-hundred-gallon storage tank. The next morning it is drawn from the tank into whatever jars or tins I have arranged for.

Thus everything is incorporated into a smooth operation with a minimum of separate steps, from uncapping combs at one end to filling jars and tins at the other, the blocks of beeswax meanwhile accumulating as a byproduct and requiring no separate operation. The whole thing takes a bit of time to set up and get into proper adjustment, its various parts harmonizing with each other and the motions of the operator, but it is well worth it.

Honey spinning is long and tiring work, as I uncap the combs one by one and start the honey through the system, and sometimes it seems endless. Yet it is the object of the whole enterprise and yields a sense of deep fulfillment when, at the very end, I uncap the last comb and roll the last stack of supers into storage, concluding another season.

EIGHTEEN

People

I have discovered a curious fact about beekeeping—if anyone pursues the craft then, in the eyes of his fellows, he forthwith ceases to pursue any other. No one seems able to picture a beekeeper in any other role. Perhaps the strangeness of the image in their minds, the image of someone keeping company with bees, simply overwhelms every other possible association.

I am a beekeeper, but I have in fact had other interests along the way, and even a few accomplishments. Yet these, though they may be known, are never the subject of comment when I enter a gathering. The question is always, "How are the bees?" There is nothing else to ask. I do not know exactly how it should be answered, if put seriously. What can one say? I do have an answer, intended more to halt further inquiry than to satisfy curiosity: "They're waxing." But it usually does not halt further inquiry. The questions that follow are likely to be more point-less than the first, even when concocted in the minds of persons learned, witty and scholarly. The next question, for instance, is usually, "Do you ever get bit?" To which the answer is, of course, that I never have yet. Sometimes it is worth adding that I receive my share of stings, but so far no bites. But usually I prefer to add no qualification that might remove the look of incredulity from my questioner's face.

Another question constantly asked, for reasons I have never fathomed, is, "What do the bees do in the winter?" And the answer to this is that they languish, like myself. And that seems to me not a purely flippant response. It fairly well describes what the bees actually do. A similarly common question of the same order is, "Who looks after the bees in the winter?" I could say "no one," but the more positive response is "the Almighty." It seems somehow to put everything in a better perspective.

Sometimes serious questions are no easier to answer. I do not know what to say, for example, in response to the question of whether my honey is "organic." Any answer that does not include a disquisition on the foraging behavior of bees is almost certain to mislead. The same is true with respect to the question of whether my honey has been "cooked." If I restore granulated honey to liquid by warming it, do I thereby cook it? The question arises from the practice of some honey-packing companies of adding the word "uncooked" to their labels—which makes as much sense as describing bananas as "boneless."

A beekeeper friend of mine was once asked how often a bee dies. This was evidently a way of asking how long a bee lives, which is a perfectly good question. But upon hearing this question, my friend was so seized by giggles that he had to withdraw and pull himself together before emerging to reply: "Just once!" It is seldom that I collect a swarm without having some bystander ask how many bees are in it. I generally produce an arbitrary answer to this—something like "seventeen thousand three hundred and sixty-two." Strangely enough, I have never been asked how I know. One of the more memorable questions I've been asked came from a lady who had for years been putting up with a nest of honeybees in the siding of her house. Her question was whether she might be able to rid herself of them by putting a dish of honey out in the yard and then, when all the bees were out drinking honey, plugging their entrance behind them. Perhaps one needs to be a beekeeper to appreciate this naiveté.

It remained, however, for my sister-in-law to ask the greatest bee question of all time. She asked me (I swear), "How much do you make on your honey business, *per bee?*" Now I suppose I could take an estimated average population of a hive, multiply this figure by the number of colonies I have, divide this into my net profit for a year (or would it be the other way around?) and get some sort of figure having maybe fifteen or twenty zeroes following the decimal point. But I do not think I ever will. I somehow feel that my life was made richer by my having received such a question, reminding me that the ways of men are sometimes, like those of God, wondrous indeed.

People, in addition to being a fertile source of questions, are also the consumers of my honey crops, so I could hardly do without them. Nor would I wish to, for I have formed many friendships among this class. For years I sold virtually my entire honey crop by mail, and even now I have customers I have never seen but whom I have come to know well from years of this association. I know their idiosyncrasies and preferences. One, for example, who orders huge quantities of honey, eats nothing but raw food. He therefore telephones each year for my renewed assurance that the honey I send him has not been warmed at all at any stage. He is even unhappy that my uncapping knife is warmed, although I have persuaded him to live with this fact.

Another of my customers, formerly a stock broker, has gained fame by achieving total economic independence for himself, his wife, and his two children through tilling a mere five acres. I have never met him, but I read about him from time to time, and we correspond each fall when he orders the year's supply of honey. Another customer is a musician and composer of fame who does his vast Christmas shopping each year by telephoning to me a long list of names and addresses.

For years honey customers were driving past my door by the hundreds, and I was not bothering to sell them honey because I did not know they were honey customers. The light began to dawn when the youngster across the road set up a card table in his driveway to offer surplus produce from his garden and, before long, from mine. Eventually, jars of my honey were set out, and I was astonished by the result. Here, it finally dawned on me, was a good outlet for honey.

Scrounging some sawhorses, wide boards and scrap lumber, and half of a Ping-Pong table for a roof, I was on my way to assembling a honey stand. My only purchases were four lengths of pipe and flanges to support the roof, a can of paint, and a tarpaulin. Altogether I spent $12.09 on this crude but serviceable stand. With the other half of the Ping-Pong table I made a roadside sign, then found another large scrap to make a second one. Using the honor system of letting people leave payment and make their own change, I stocked the stand and awaited results.

No sooner did I get the first roadside sign erected than the next driver stopped for a purchase, while I watched with interest from behind the shrubs. He will never know that he was the first of a long line of happy customers. In no time my dear wife was busy keeping the stand supplied and the honor box emptied. Since then the business has grown steadily as old customers have returned and new ones have been added. The honor system still prevails, with seldom any significant discrepancies.

The stand has also been a rich source of friendships and of a unique status for its proprietor. Upon entering many gatherings I need only to mention the honey stand to be instantly identified, even by persons living some distance away. The roadside

signs, which have grown in both number and size, mark my location in the world. Four large colleges are fairly near by, so my stand draws great numbers of students who offer me their assessments of the world's ills. Somehow they seem sure that anyone who raises bees will certainly share their appraisals of the world and its problems and they are almost entirely correct in this. Some have dreams of keeping bees themselves, a vocation that evidently seems to them in keeping with the nobility of their thinking, which they hope to sustain through their lives. It is my own fervent hope that they will sustain these ideals, and perhaps in some cases unite them with beekeeping.

Not everyone can be a beekeeper. The tiny but pesky sting will always keep the membership in this strange class to a proper number. But for one who can see beyond this, it is, indeed, an enviable life, opening one's eyes not only to nature, to philosophy, to that life of the spirit that is basic to religion, but also to the warmth and idealism that dwells in so many of the human beings who are brought within one's association.

NINETEEN

Keeping Bees for a Living

Can one make a living keeping bees? The question has always intrigued me. More than once I have been tempted to forsake everything else to prove it in the affirmative. Perhaps I shall yet. What I have done is to compromise, by neglecting my other gainful pursuits while not entirely repudiating them, for actually I love those pursuits too. It is not a perfect solution, but few things are. The bees have seldom been far from the center of my interests and, as I get older, they become more central.

But *can* one actually make a living keeping bees? Many have done it, many do it now, but the answer is not really as simple as those facts suggest. It depends on what one means by a living, what kind of person one is and, above all, how good a beekeeper one is. In my own thinking on this question I have decided that the ancient rule of the good life—"neither too much nor too little"—applies rather well here. Many of the rewards of beekeeping are intangible, and they are easily lost if one is dedicated single-mindedly to gain. Indeed, one can go further and say that, if one's sole aim is a livelihood, then almost any of the familiar ways of gaining it is better than beekeeping.

It is not the proper spirit for beekeeping to begin with, and one had better turn to something simpler, easier and less at the mercy of capricious weather.

If, on the other hand, you can be content with a simple and modest life and with treasures that are sometimes invisible to the rest of humanity, and if, in addition, you are resourceful and have a sense for nature, then you can indeed make a modest living from beekeeping. At the same time you will have a very good chance at that elusive mode of existence that the philosophers try to define as happiness. It can be a good life and a self-subsistent one. It cannot be one of riches, nor one of leisure, yet its non-negotiable riches exceed those of any work I know.

Many persons see in beekeeping a possible fulfillment of their desire for simplicity and a life close to the earth. This is especially true of young people who are not inspired by the goals that lured their parents. They swarm to my honey stand and pile question upon question. Sometimes I have a chance to take them around to my yards with me. I am glad for the occasional help and the company of those who love nature as I do, and I do not care if some of them are a little sloppy. Anyone can put on a necktie.

It must be emphasized that in apiculture, as in the faith, many are called but few chosen. The world is full of people who have tried keeping bees and have stopped. The image of the bee master quietly going about his work, the hum of the bees over his head, the open air and the sunshine, the opening flowers, the quantities of sweet honey that reward this agreeable activity—this is a pleasant image but not a complete one. One must add the occasional sting on the eyelid or nose, the tank of honey that sooner or later runs over noiselessly, the hive that falls from the truck, the exhaustion of a day of swarm gathering, the truck hopelessly stuck in the mud, the aching limbs during the harvest, the numberless frustrations that, while they reinforce the determination of the few who were born to be beekeepers, totally defeat the vast majority who were not. All the skill in the world is of little use without the fortitude to cope with discouragement and the frequent threat of disaster. Beyond this is the plain fact that many lack even the skill or the temperament ever to acquire it.

I thought of this one day after I had been approached by a young man who told me he had just purchased several hives of bees and was now living in a tent not far away. I soon concluded that he was not among those chosen. As I worked with my bees he withdrew to a safe distance, his hands, bare like mine, safely hidden behind his back. He shortly left to get his bee veil and returned wearing it backward. He had no comprehension of what I was doing and was unwilling to approach close enough to find out. I never saw him again, but I doubt that he ever became a beekeeper.

Spring never arrives without someone asking me how he might go about becoming a beekeeper. Something about the season, of course, fuels enthusiasm. I encourage everyone I can, but most do not get beyond the initial steps of thinking and dreaming. It is just as well that they do not; it is the person who dreams of bees in the winter who has a chance of becoming a beekeeper.

Perhaps no agricultural pursuit is as flexible as apiculture or requires less of a permanent investment. Someone who produces maple syrup has to take what nature has provided; he cannot go out and buy up more maple trees and truck them around. But the beekeeper can buy up colonies of bees and cart them more or less where he pleases or, when the occasion arises, he can sell them, all or part. He does not need to own the land on which he puts them, and the bees forage over thousands of acres that he does not till or tend or pay taxes on.

The ease with which a beekeeping enterprise can be expanded has lured some into enormous holdings. There are beekeepers who own thousands of colonies spread over great areas. Their honey houses are large complexes of extracting machinery and their crops are sold in steel drums. But when I see the big commercial beekeepers with their heavy trucks and hoists and equipment I certainly do not envy them. It is no picture of a simple life. It is also a hard way to make a living, loaded with headaches. For me it would be self-defeating.

Apiculture is about the only branch of husbandry that the drive for size has not robbed of its joys. The modern poultry farm is nothing but an egg factory, bearing no resemblance to the chicken yards we knew as children, with their clucking and daybreak crowing, which we no longer hear. Perhaps I am too much given to nostalgia, but I sometimes wonder whether today's children will have much to look back on. The family farm has been replaced by huge commercial operations that have lost the heart and soul of farming. The baneful apparatus of the factory now spreads itself over the rural landscape, knocking down everything in its path in response to the demand for increased production, often just in response to greed. It is becoming rare to see farm buildings that are still used. Some houses and barns are kept up, but most are collapsing and others are taken over by people who spend the better part of their lives in the cities. The scene fills one with sadness.

Beekeeping, however, can still be carried on in the traditional manner, as a backlot or sideline operation, or even as the means to a modest but wholesome livelihood. It has the same charm and fascination as in the days of old. It has not changed much in the course of a century. It seems to be an almost incorruptible craft.

The bees themselves are no different from those our forefathers kept nor, indeed, from the inhabitants of the nearest bee tree. For decades specialists have been trying to breed superior strains of bees, to bring this tiny animal under the same domination to which other livestock has been subjected, but with only limited success. They have even artificially inseminated queens, but this too has had only limited significance. Bees revert to nature after only one generation, for the queens then mate high in the air and out of human control with whatever drones overtake them. I personally hope the bees will always win this tug of war, that they will never be subdued and corrupted by the human drive for greater production, that they will never become the stupid and denatured animals that chickens and pigs have become through domestication. When I see a bee tree I know its inhabitants are the evolutionary product of millions of years, and that what I call "my own" bees are but the smallest step from the bee tree. The forests still lure them back and always will. The bees go on from season to season, just as they have for millions of years, and as they did before the human race appeared on the earth. I find this thought deeply satisfying.

Many men and women have a deep, instinctive pull to the land and husbandry. It is no longer satisfied by modern methods of commercial farming, and not many persons in touch with reality are tempted to try earning a livelihood in gardening, however satisfying this may be in other ways. For some people, however, beekeeping can fill this need completely. With only a minimum of tools and investment, augmented by resourcefulness in the use of what comes to hand, a beekeeper can play his part in the great cycles of nature, find a satisfaction that few pursuits offer, augment his livelihood or even gain the whole of it from his bees.

The beginning can be made with a single colony, set down almost any place, even on a roof or porch. Unlike other livestock, bees do not need to be fed except under infrequent and unusual circumstances. There is no butchering or cleaning of pens, disposing of waste or refuse, or other constant chores ordinarily associated with husbandry. One can come from the bee yard straight to the dinner table with no smell of the barn-yard. There is no despoilment of the landscape or exploitation of the earth's resources. On the contrary, the bees make an essential contribution to ecological well-being. Equipment can be kept simple and much of it can be made up with only a few common tools even by persons who, like myself, have little talent in this area.

My own formula, so far as the size of my bee operation is concerned, is to draw the line at that point where I would need to hire help. That is where the headaches begin and the law of diminishing returns begins to be felt. It is the line that marks the difference between two quite different pursuits and two kinds of beekeeper. If one crosses that line, if one depends upon paid help, even though it may be seasonal and unskilled, then there is little choice but to increase holdings as much as strength and circumstances permit, just to offset the increased overhead that hired help represents. This, of course, sooner or later means the hiring of more help and the purchase of more heavy equipment. Economic return becomes the governing consideration once one has chosen this manner of commercial beekeeping. Having become big, one finds that the thrust of activity is to become even bigger.

As a one-person operation, on the other hand, or sometimes as a joint endeavor of husband and wife, beekeeping can be totally fulfilling. Every bee yard is a little kingdom and the honey house a little castle. One does not get rich at this life, but it can help substantially in the domestic economy. One can sell the honey directly, at the top price, instead of dealing with processors and taking what they are willing to offer. The beekeeper can take a special pride in his work and in its product, which will bear his own label.

How large, then, should such a beekeeping enterprise be? Where is that point at which one needs to hire help? Can one make a living with bees, just on his own?

This depends on many things. Someone who relies on other sources for his basic livelihood and will continue to do so—the teacher, the mail carrier, whatever—and who looks forward to a retirement income can become a sideline beekeeper, keeping perhaps one or two hundred hives in several locations. He will not need to hire help and his bees will be profitable, provided he has the qualities needed for keeping them.

But now let us consider something more definite. Consider a strong, intelligent and resourceful person who has the temperament and qualities that make an outstanding beekeeper. Can such a person, perhaps in concert with a willing wife or husband, make a living from bees?

I shall show that he can, but I shall need a few assumptions, which, though not unrealistic, will be helpful to my case. I shall assume that this person is already an expert beekeeper and has accumulated over the years the equipment necessary for about three hundred colonies of bees. I shall disregard how he came by this equipment or how it was paid for. I shall also assume that he has a small truck or good car and a fair-sized utility trailer. Finally, I shall assume that he has an adequate honey house equipped with extractor, tanks and the rest.

It should be noted that I am not assuming that our beekeeper is going to have to finance the cost of all this equipment. I assume he has already done so, or that in any case he somehow now owns it. And the question is simply whether he can, thus equipped, make his whole living with it.

We now need only two more assumptions. These are, first, that our beekeeper lives in a primary honey region and that he lives near cities or near a well-traveled road where perhaps a third of his crop can be marketed at retail.

This person could manage three hundred colonies single-handedly. He would be busy in the spring and fall, but at other times the work would not be heavy. In fact, he would have little to do through the winter except putter. His honey crop would average about twenty-five or thirty thousand pounds per year in a decent beekeeping area, sometimes less, sometimes more, depending upon variations in weather and other factors from one season to another, as well as upon the beekeeper's skill. About a fourth to a third of this crop could be sold at retail, assuming a beekeeper well-located—for example, one who lives on a well-travelled road or who can sell directly to tourists—as well as a beekeeper who is resourceful and imaginative in his approach to selling.

It is not easy to translate these generalities into figures, because the price of honey, like everything else, changes, and estimates can be rendered obsolete rather quickly by the passage of time. Currently, with honey selling from the grocer's shelf for about a dollar and a half per pound, a beekeeper can net about a doller per pound for honey sold at retail and about half that for a crop marketed at wholesale, tripling these figures for comb honey of good quality. This would translate into an annual income of fifteen to twenty thousand dollars in today's economy. To get an approximate adjustment of these figures to other times or places, one can assume a wholesale price equal to

about a third of what he finds honey selling for in the average grocery, and a net retail price equal to about two-thirds of that figure.

There would, needless to say, be costs, for fuel and supplies, which would reduce the foregoing income estimates by perhaps ten or fifteen per cent. But, most significantly, there would be no labor costs. What would then be left for the beekeeper to live on would not make him rich, but it would at the same time be far from insignificant.

It must be pointed out that no account has been taken of investment in equipment, so the summary given is meaningless from the standpoint of standard methods of accounting. It is nevertheless meaningful from the standpoint of common sense, for a beekeeper can accumulate, and pay for, the equipment over time while depending upon some other source for his livelihood.

The pursuit of beekeeping, whether as a source of livelihood or, as is usually more practicable, as a sideline, is totally engrossing. It offers fulfillment for the golden years whose approach seems so relentless and filled with emptiness to those who have somehow become estranged from nature. Sometimes the world seems on the verge of insanity, and one wonders what limit there can be to greed, aggression, deception and the thirst for power or fame. When reflections of this sort threaten one's serenity one can be glad for the bees, for the riches they yield to the spirit of those who love nature and feel their kinship with everything that creeps and swims and flies, that spins and builds, to all living things that arise and perish, to the whole of creation of which we are only a part, like the bees.

CPSIA information can be obtained
at www.ICGtesting.com
Printed in the USA
BVHW081949241119
564688BV00004B/162/P